DEVELOPING TABLET PC APPLICATIONS

CLAYTON E. CROOKS II

CHARLES RIVER MEDIA, INC.
Hingham, Massachusetts

Editor: David Pallai
Production: Publishers' Design and Production Services, Inc.
Cover Design: The Printed Image

CHARLES RIVER MEDIA, INC.
10 Downer Avenue
Hingham, Massachusetts 02043
781-740-0400
781-740-8816 (FAX)
info@charlesriver.com
www.charlesriver.com

This book is printed on acid-free paper.

Clayton E. Crooks, II. *Developing Tablet PC Applications.*
ISBN: 1-58450-252-5

Library of Congress Cataloging-in-Publication Data
Crooks, Clayton E.
    Developing tablet pc applications / Clayton Crooks.
        p. cm.
    ISBN 1-58450-252-5 (Pbk. with CD-ROM : alk. paper)
1. Pen-based computers.   2. Application software—Development.
3. Microsoft Visual BASIC.   4. BASIC (Computer program language)
5. Microsoft .NET.   I. Title.
    QA76.89.C76 2003
    005.265—dc22
            2003017815

Printed in the United States of America
03 7 6 5 4 3 2 First Edition

CHARLES RIVER MEDIA titles are available for site license or bulk purchase by institutions, user groups, corporations, etc. For additional information, please contact the Special Sales Department at 781-740-0400.

DEVELOPING TABLET
PC APPLICATIONS

This book is dedicated to my wife Amy,
who provides never-ending love and support.

Contents

Preface

The Tablet PC OS is the newest Microsoft OS and provides a unique set of features for users and developers. The pen and ink are the two most important additions, but they are certainly not the only important aspects of Tablet PC development. That's where this book differs from the rest of those on the market. Along with the pen and ink, we develop applications that take advantage of some of the other popular features of the Tablet PC, including power management, 3D rendering, and games—to name a few. The following list details some of the wide range of programs we'll develop:

Full screen utility: Introduces some key concepts for pen and ink

Screen rotation: Rotates the screen programmatically

MP3 player: Plays music with the Tablet PC

Media player controlled with gestures: Uses gestures to control a multimedia player

Speech input with SAPI: Dictates using SAPI

Speech input with Agent: Allows command and control using Agent

Power management: Directs the power management from within your own applications

Pong: Includes unique options for controlling players with the pen; one of the games that we develop

Faxing: Faxes from your own programs

3D rendering: Renders 3D scenes using OpenGL, DirectX® 9, and third-party engines

Many more . . .

Acknowledgments

There are many people who have been involved with the development of this book, and because of their hard work and dedication, you are now holding it. First, I'd like to thank everyone at Charles River Media, and especially David Pallai, for the opportunity to write another book. It continues to be a great pleasure to work with you. To say that Karen Annett and Bryan Davidson, who were given the enormous task of coordinating and editing the book, went above and beyond normal expectations is an understatement. Thanks for all of the help and input. Lastly, I'd also like to thank my family and friends for their help and support during the long hours it took to write this book.

1 The Tablet PC

This book is based on developing applications for the newly released Tablet PC using Visual Basic® .NET™. In this chapter, we focus on the specifics of the Tablet PC, including what makes a Tablet PC unique and how this affects the development process. In subsequent chapters, we learn about general VB .NET development, and after building a solid foundation, we focus on developing applications specific to a Tablet PC. In doing so, we take advantage of the Tablet PC hardware, which brings us to the first concept—the requirements of a Tablet PC.

WHAT IS A TABLET PC?

As we create applications for the Tablet PC, we will encounter many things that are already common to you, such as hard drive storage, memory restrictions, and a display. However, there are many things that are specific to a Tablet PC, and we need to keep them in mind as we develop applications. Some of the hardware specifications enable us to create applications that will take advantage of the new technologies. At the same time, we also have to deal with some of its limiting factors. Keeping track of both is imperative to a Tablet PC project.

First, a Tablet PC has many of the same hardware and functions that are available on a regular desktop or notebook PC. It runs what is essentially Windows® XP as an operating system, which allows the vast amount of existing Windows® XP-based applications to be used. Common programs, such as Microsoft Word and Excel, work very well on a Tablet PC, as do most applications that work on Windows XP-based systems.

Because many of the features and concepts are identical, it might benefit us to look at three devices that detail what a Tablet PC is not. A Tablet PC is not a:

■ **PDA:** PalmPilots™ and Pocket PCs, two of the most popular variations of Personal Digital Assitants (PDAs), have a few characteristics that are shared with a Tablet PC. They both use a stylus for input on a built-in screen, although this is where the similarity ends. PDAs have severely limited processing power and a very small screen, usually measuring a maximum of a few inches in width and height. On the other hand, a Tablet PC has a much larger screen with some models offering up to 12.1" at the time of this writing. PDAs typically do not have hard drives and offer very small amounts of RAM, whereas Tablet PCs have large hard drives up to 60GB and may offer 1GB of RAM in a single machine. Power is definitely what separates the PDA from a Tablet PC.

■ **Notebook:** Although many features are similar, a notebook computer does have some very specific differences from its Tablet PC counterparts. They are each mobile devices that can rely on battery power. They have similarly sized screens, although notebooks tend to have a slightly larger offering in this area. They can both offer a standard set of ports, such as Universal Serial Bus (USB) and Video Graphics Array (VGA). You will notice some ports missing from a Tablet PC, such as a serial port, game port, or PS/2 port. These ports are missing because a Tablet PC only supports plug and play capable ports and these ports do not support plug and play. Additionally, both notebooks and Tablet PCs typically offer keyboards to be used for data entry. Notebooks have the keyboards built-in, whereas many Tablet PCs require external keyboards or docking stations. A Tablet PC is much different in that a stylus pen is often the preferred method of data entry, whereas a notebook does not offer this option at all. Also, the screen for a Tablet PC offers both landscape and portrait modes, something lacking on a notebook screen. Tablet PCs are also typically much smaller than all but the smallest of subnotebooks.

■ **Book readers:** Recently, there has been a great push in the media hyping technologies for reading electronic books. A Tablet PC can be used for such purposes with software such as the Microsoft Reader and Adobe's Acrobat, so in some ways, it is similar to an electronic book reader. This is where the comparisons end, however, because a typical electronic book reader offers only the built-in functionality, whereas a Tablet PC can run many different types of software.

Now that we know what a Tablet PC is not, we can more easily look at what makes a Tablet PC. These are typically the ideas that make up a Tablet PC and separate it from any of the other devices we have mentioned:

■ **Pen input:** Tablet PCs, as we have already mentioned when looking at the PDA comparisons, have a digitizer screen that accepts pen input. When not being used for digitizing input, a pen offers the functions typically offered by a mouse.

■ **Digital ink:** When you write on a Tablet PC, you are presented with data in real time as you use a pen. The data, in the form of digital ink, is one of the unique features of the Tablet PC and one that we focus on in greater detail later in this chapter in the Digital Ink and the Digitizer section and throughout the book.

- **Speech input:** Built-in to a Tablet PC is the ability to use speech input. This can be used for dictating content or controlling applications.

- **Extended battery life:** Power management is a big focus of the Tablet PC as is the ability to resume quickly from standby mode. In standby mode, a Tablet PC must have the ability to maintain its charge for 72 hours and also resume when needed in a minimum of 2 seconds. These are the published requirements made available by Microsoft.

- **Hardware buttons:** Many Tablet PCs offer programmable buttons. This is not a requirement, and many hardware manufacturers offer variations, so it is something that needs to be looked at carefully. One way to handle this is to handle the inputs from the hardware buttons like you would input from the keyboard, allowing the user to change the buttons if needed. We look at handling hardware button presses in Chapter 15, Tablet PC Screen Rotation and Special Buttons.

To summarize for our needs, a Tablet PC is a small and multipurpose computing device that uses a pen (or stylus) for input and runs Windows XP Tablet PC Edition.

IMPORTANT CONCEPTS FOR TABLET PC DEVELOPMENT

Now that you know, at least by a simple definition, what makes a Tablet PC unique, we'll take a look at how its features are implemented. We look at the Tablet PC Input Panel, digital ink, and the digitizer/screen. More importantly, we look at how they affect a development project.

Tablet PC Input Panel

The Tablet PC Input Panel (see Figures 1.1 to 1.3) is the general method for speech and pen input for a Tablet PC. If you use a Tablet PC with standard applications, you will soon realize that it is indispensable. This is not to say that a developer writing new applications should rely on the Input Panel as a method of input or as a crutch to be sloppy with a design. You would, however, be advised to borrow ideas from it, such as its size in relation to the total available screen and its ability to hide from view until needed. Think of the Input Panel as it was intended—a general input area and not as a way to enter data into an application specifically designed for a Tablet PC.

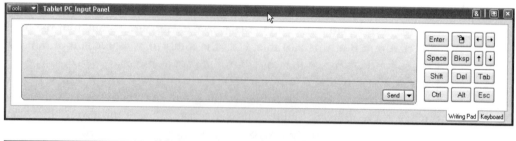

FIGURES 1.1 TO 1.3   The Tablet PC Input Panel in various states.

Digital Ink and the Digitizer

The digital ink used in Windows XP Tablet PC Edition seems to be a very obvious thing to mention, but there is more to the ink than an image file. Instead, the ink is stored as a series of mathematical equations, known as Bezier curves. This allows the file sizes needed to store digital ink to be small and very efficient. Tablet PCs utilize an electromagnetic digitizer, which accepts input from a stylus that contains an electromagnetic coil. Although the digitizing process is very technical, it can be summed up in the following steps:

1. The operating system (OS) captures pen motion coordinates and renders it as ink.
2. The pen strokes are sent to a recognizer.
3. The OS displays an interpretation of writing.

There is a great deal of technical information that could be presented on this topic, but because the information is not really necessary for development purposes, we won't go into it in any detail. The Microsoft Web site, as well as Web sites from hardware manufacturers, contains more detailed information about the hardware and technical aspects of the digitizer.

While we're discussing the digitizer, it seems like a good idea to also look at the screen and how it affects development. Because much of the input of an end user is pen-based, it is imperative to consider this as you design an application. Although writing on the screen seems fairly intuitive and natural, the process can cause some difficulty for the user. For example, one very big problem is the way menus are displayed when they are pressed with a pen. To explain this, refer to Figure 1.4, which displays a mouse pointer choosing a menu item in Microsoft Word. If you look closely at the text, you can see that this is the Insert menu as I'm editing this chapter inside Word. Clicking on the menu doesn't cause any problems as the small mouse pointer is the only thing in the way. Now, look at Figure 1.5. This displays the same menu as it would appear to a right-handed user clicking on it with a pen. You can see that most, if not all, of the menu is hidden by the user's own hand.

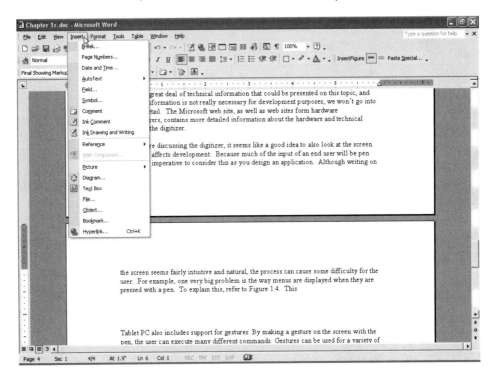

FIGURE 1.4 A menu being selected with a mouse.

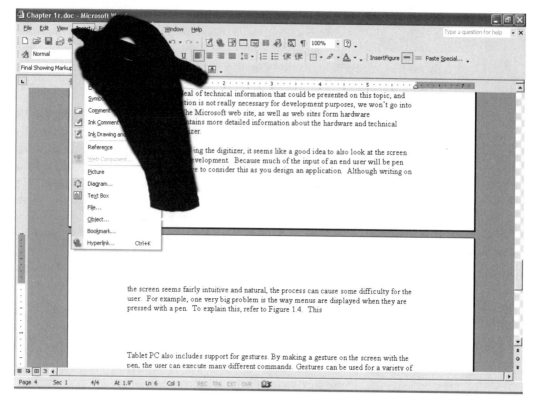

FIGURE 1.5   The same menu being selected by a pen is hidden by a hand.

Many times, as an end user, you can move your hand to the side to view the information that is being displayed in a hidden menu. There are times, however, when this is not practical, and in general, needing to constantly move to view items can cause considerable delays and frustrations for an application user. There are also times when moving just won't work, such as a ToolTip, which requires the mouse pointer to be in a particular location in order for the ToolTip to be displayed. Further compounding this problem is the fact that some users are also left-handed. This is a problem that should be remembered, but unfortunately, you will never completely avoid it. That being said, Microsoft has given us the ability to set up the handedness of the end user. This allows the user to position their menus away from the pen. You can set this in the Tablet PC Control Panel (see Figure 1.6). When you click the Tablet and Pen Settings icon, you will see the Tablet and Pen Settings window displayed, as shown in Figure 1.7. This is where you can select the handedness settings.

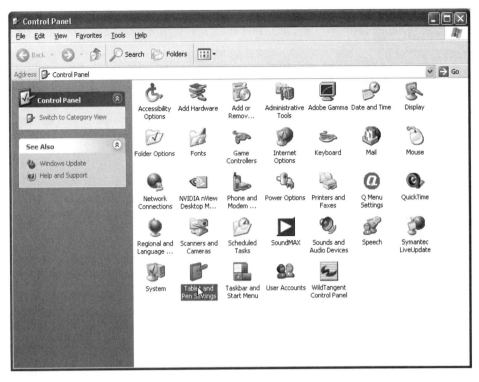

FIGURE 1.6 The Tablet PC Control Panel.

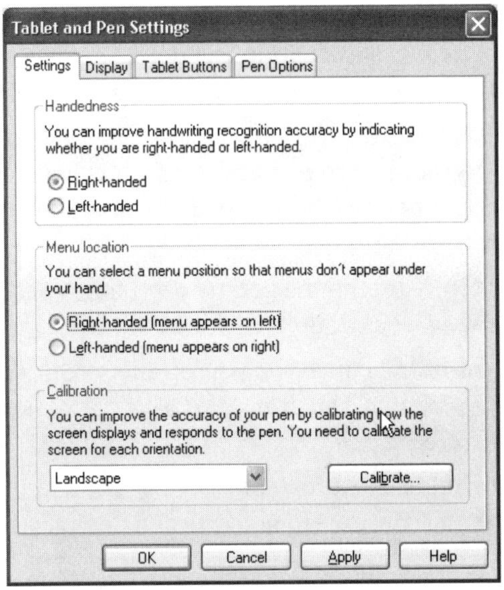

FIGURE 1.7 Settings for the pen.

The handedness settings only affect ToolTips and standard menus. If an application has been developed with a custom menu, it will not work with these settings.

An additional consideration related to the Tablet PC screen is that the screens support both landscape and portrait modes. You will need to allow your interface to change as the screen changes so that the end user can use both modes. Generally speaking, if your application is going to be mostly pen-based, it is probably safe to assume that the user will use their Tablet PC in portrait mode, like a sheet of paper. This is not completely accurate though, so you should do your best to allow both modes.

Types of Ink

Among the most important things you'll have to decide for a pen-enabled application is how you will use the ink in your application. That is, will the ink be used simply as ink, such as the writing on a piece of paper, or will it be converted or changed in some other way? In other words, are you planning to capture the ink as handwriting, or are you going to convert it to text, graphics, or manipulate it in another manner? Determining which type of ink you need is dependent on the type of application you are going to design.

Pen Interactions

Another key ingredient for a Tablet PC is to determine the way in which the pen will interact with your application. There are a few common tasks that a pen can do:

Hover: The pen stays within the detection range of the digitizer but does not actually touch the screen. This is similar to a standard mouse moving on the screen and the traditional mouse arrow also moving to indicate this.

Tap: If you touch the pen and then quickly lift it from the screen, it is a tap. This is the same as a left-click of a mouse and is treated as such in Windows applications.

Double tap: Tapping the pen twice in quick succession is called a double tap. This is the same as double-clicking a mouse.

Press-and-hold: Touch the pen on the screen and hold it in a single location for a few seconds. This functions as a right-click of a mouse when you lift it from the screen.

Hold-through: The press-and-hold goes hand in hand with the hold-through action. If you continue to hold the pen down even longer than the requirement for the press-and-hold, instead of getting a right-click action, you get the same action as holding down the left mouse button.

Drag: Touch the screen with the pen, and while holding down, move a distance on the screen, eventually lifting the pen from the screen. This is the same as dragging a mouse.

Hold-drag: Similar to a drag, you can do a press-and-hold and then drag to simulate a right-mouse-button drag.

Gesture: A gesture is a unique method of input with a pen. By drawing a certain pattern, the Tablet PC recognizes it and performs an action based on it. Gestures can be used for a variety of common actions or commands similar to a keyboard shortcut.

SUMMARY

In this chapter, we covered the basics of the Tablet PC. We began by comparing its features to those of other computing devices and came up with a simple definition for a Tablet PC. Next, we briefly looked at how the Tablet PC can affect the development of an application. In the next chapter, Introduction to Visual Basic .NET, we begin to look at Visual Basic .NET, the language which is used in this book. We spend several chapters covering the basics of VB .NET and return to specifics related to the Tablet PC in Chapter 12, Obtaining the Tablet PC SDK. If you have experience with VB .NET, you can quickly skim these chapters. Otherwise, these chapters lay a good foundation that you can use to build Tablet PC applications.

2 Introduction to Visual Basic .NET

Although this book is based around developing projects for the Tablet PC, it is important to have at least a basic understanding of the Visual Basic .NET development environment and programming language. As a result, this chapter introduces you to Visual Basic and guides you through the creation of your first Visual Basic program. Topics that you'll be exposed to include the basics of the Visual Basic Integrated Development Environment (IDE), basic language, variables, and the built-in components.

 We also look at the difference between VB .NET and earlier versions, such as Visual Basic 6 in Chapter 3, Working with VB .NET.

For those of you already well-versed in VB .NET or earlier versions of Visual Basic, you will find that some of this chapter deals with topics with which you are undoubtedly familiar. Although most of the concepts have remained the same, the IDE has been changed, and there are many new ideas we touch upon in this opening chapter. If you are a beginner or intermediate programmer and are new to VB .NET, you will find this chapter helps you build a solid foundation onto which you can base your future Visual Basic learning.

HISTORY OF VISUAL BASIC

Visual Basic has been in existence for over 10 years after debuting in the spring of 1992. It received a tremendous amount of press coverage, which propelled it into what is now the most widely used programming environment in the world. The first couple of versions were very simple and probably weren't useful for much more than prototyping an application you planned to write with something else. In subsequent releases, beginning with the database connectivity in VB3, Microsoft began to add features that transformed the tool that many considered a toy into a very usable product. In version 4, there was a limited ability to create objects and with versions 5 and 6, came additional object-oriented features. Now, with VB .NET, Visual Basic has been altered yet again and with these changes comes a shift in the way you'll develop applications. We'll look further at the language changes and how to upgrade VB6 projects to .NET in Chapter 3, Working with VB .NET, but for now, we'll introduce you to the VB IDE.

THE INTEGRATED DEVELOPMENT ENVIRONMENT

We begin by looking at the VB IDE, which is one of the reasons why it is such a popular tool. It provides everything you need to develop applications in an easy-to-use and learn Graphical User Interface (GUI; pronounced Gooey).

Like many Windows applications, Visual Basic has several ways in which it can be opened. First, and probably the easiest, is accessing it through the Start menu—the exact path required to access this shortcut is dependent upon your installation and may differ on individual machines. Another option is to create a shortcut on your desktop, which executes Visual Basic when you double-click on it. There are several additional options such as double-clicking on files that are associated with the environment, so you should simply use the option that best suits your individual needs.

If you have experience with previous versions of Visual Basic, the first thing you'll notice is the lack of the familiar New Project window being displayed on startup. Instead, you are presented with a new Start page that appears very much like Figure 2.1.

The Start page is displayed at first startup showing the My Profile option, which allows you to set the environment so it reflects your personal settings. The customization is an important option because it can help reduce your learning curve and make your job much easier. You'll notice that there are three settings you can change in the Profile portion of the page: Keyboard Scheme, Window Layout, and Help Filter. Because we're using Visual Basic, you can set the Profile drop-down to read Visual Basic Developer. When you do so, it makes changes to the other options by automatically setting the options to Visual Basic preferences. You can see the changes reflected in Figure 2.2.

There are several other interesting items on the Start page that are new to this version of Visual Basic. First, there is the Headlines option, which pulls information from the Internet for you and displays it inside the environment. You can see an example of the Headlines tab in Figure 2.3. The information in this tab consists of a variety of information, such as features of a new OS, technical articles and white papers that give specific details for topics, and a knowledge base that allows you to

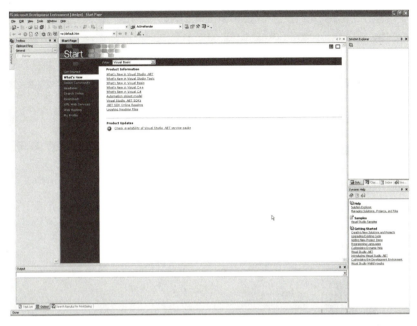

FIGURE 2.1 The Start page as displayed at startup.

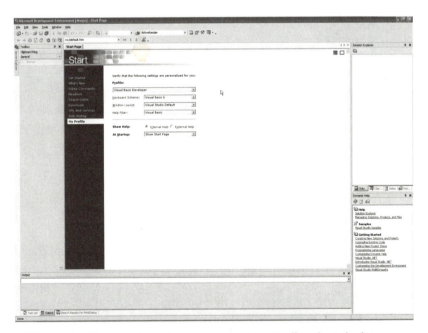

FIGURE 2.2 Visual Basic preferences are automatically selected when you set the profile.

search through thousands of topics. Having this information available directly in the IDE is a tremendous time-saver.

FIGURE 2.3   Headlines are retrieved from the Internet and displayed in the IDE.

The Get Started item, which can be seen in Figure 2.4, allows you to quickly view the projects you have been working on in a very convenient format. You can open any of the files listed in the Project tab by simply clicking on it. You can also start a new project by clicking the New Project button.

For most of the projects in this book, you will be developing Standard Windows executable files. In previous versions, you would have selected Standard EXE project from the Project window. The new VB .NET Project window, which can be seen in Figure 2.5, has several new options, including Windows Application, which is similar to the Standard EXE project in VB6. When you create the project, you are required to give it a name and a location in the dialog boxes that are listed beneath

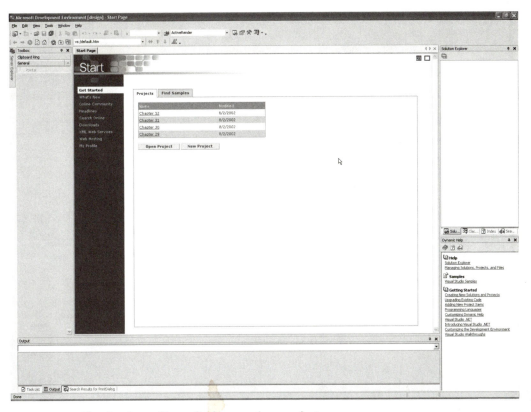

FIGURE 2.4   The Get Started item displays previous projects.

the project types. After you have set the appropriate options, you can click the OK button to create the project.

After you click OK, the VB .NET IDE displays and looks similar to Figure 2.5. One thing you may notice immediately is the change in file extensions for VB forms, which were given names with an extension .FRM in previous versions of VB. In VB .NET, items that are used in a project are given the extension .VB. Also given a new extension is a class, which also uses the .VB extension.

As you can see in Figure 2.5, the Visual Basic IDE is fundamentally a collection of menus, toolbars, and windows that come together to form the GUI. There are several main windows that appear in the default Visual Basic IDE, along with several toolbars.

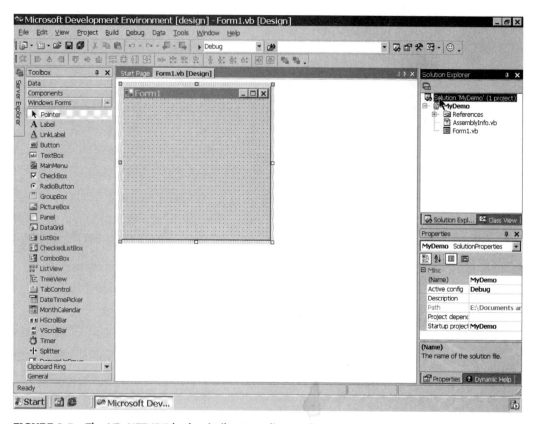

FIGURE 2.5   The VB .NET IDE looks similar to earlier versions.

THE MENUS

As you can see in Figure 2.6, the Visual Basic IDE contains a menu bar and title bar that appears very similar to most Windows applications. The title bar serves as a quick reminder of what you are doing inside the IDE. For instance, unless you have changed something, the title bar should currently read something similar to "WindowsApplication – Microsoft Visual Basic .NET [design] – Form2.vb[design]."

File   Edit   View   Tools   Window   Help

FIGURE 2.6   Menu and title bars provide information similar to most Windows programs.

The menu bar provides functions that you would expect from any standard Windows application. For instance, the File menu allows you to load and save projects, the Edit menu provides Cut, Copy, and Paste commands that are familiar to most Windows users, and the Window menu allows you to open and close windows inside the IDE. Each menu option works like any other Windows application, so they don't need any real introduction. You shouldn't be overly concerned with all of the options at this time, because we spend some time on them as we encounter them throughout the book.

THE TOOLBARS

We'll now take a few moments to look at the toolbars in the Visual Studio IDE.

The Standard Toolbar

The Standard toolbar, which is displayed in Figure 2.7, is also comparable to the vast majority of Windows applications. It provides shortcuts to many of the commonly used functions provided by Visual Basic. Along with the Standard toolbar, Microsoft has provided several additional, built-in toolbars that can make your job a little easier. To add or remove any of the toolbars, you can right-click on the menu bar, or you can select Toolbars from the View menu.

FIGURE 2.7   Toolbars provide shortcuts to many of the common functions.

Individual Toolbars

The individual toolbars include the Debug, Edit, and Form toolbars. The Debug toolbar, which is visible in Figure 2.8, is utilized for resolving errors in your program and provides shortcuts to commands that are found in the Debug menu.

FIGURE 2.8   Shortcuts in the Debug toolbar
are helpful when finding errors in your program.

The Layout Toolbar

The Layout toolbar (see Figure 2.9) includes many of the commands in the Format menu and is useful when you're arranging controls on a form's surface. For instance, you can quickly and easily align objects or center them horizontally or vertically on a form.

FIGURE 2.9 The Layout toolbar displays buttons specific to editing features.

Whether you decide to display these toolbars is purely a matter of personal taste as the functions they provide are generally available in menu options. Several factors, such as your screen size and resolution, may make their use impractical.

Custom Toolbars

You can create a custom toolbar or customize the appearance of the built-in toolbars by following a couple of steps. First, right-click on any toolbar, and then select the Customize option. From the Customize window that appears, which can be seen in Figure 2.10, click the New button and type a name for the new toolbar. The name appears in the toolbar list, and after making sure that its check box is selected, click the Commands tab, which displays a list of available menu commands. From the list of categories and commands, select the options you want to have on your toolbar. The changes are automatically saved, so continue placing options on the toolbar until you are finished, and then simply close the window. You can now use your toolbar like any other.

FIGURE 2.10 The Customize window allows you to create or customize a toolbar.

THE WINDOWS

In addition to the menus and toolbars, there are several windows that you need to become familiar with to get a basic grasp of the Visual Basic IDE. The Toolbox window displays some of the built-in Visual Basic controls. The Form Designer displays the basic building block of a typical Visual Basic application. The Code Editor is where you'll actually type the code for your application. Solution Explorer displays the objects that make up the project on which you are working, and you can position and view the forms with the Form Designer window. Lastly, you can set properties of the components and forms with the Properties window.

The Toolbox

The Toolbox, which can be seen in Figure 2.11, is probably the window that you will become familiar with the quickest because it provides access to all of the standard controls that reside within the Visual Basic runtime itself. These controls, known as intrinsic controls, can be sorted by right-clicking on the Toolbox and choosing Sort Items Alphabetically from the pop-up menu.

Pointer: Allows you to select controls that have already been placed on a form. The pointer is the only item on the Toolbox that isn't a control.

Button: Allows you to create standard buttons for input, such as OK or Cancel. Much like the TextBox, CommandButtons are used for input on almost every frame.

CheckBox: Allows the user to select True/False or Yes/No.

CheckedListBox: Displays a list of items that can display a check mark next to items in the list.

Color Dialog: Allows the user to choose colors.

ComboBox: Contains a list of items, but only provides support for a single user selection. This is similar to a ListBox control.

ContextMenu: Gives users access to frequently used menu commands.

CrystalReportViewer: Allows a Crystal Report to be viewed in an application.

DataGrid: Displays data in a tabular view.

DateTimePicker: Allows the user to choose a single item from date/time.

DomainUpDown: Displays and sets a text string from a list of choices.

ErrorProvider: Allows you to show the end user something is wrong.

FontDialog: Displays a dialog box for fonts.

FIGURE 2.11   Standard controls as well as
ActiveX controls are displayed in the Toolbox.

GroupBox: Provides grouping for other controls.

HelpProvider: Allows you to associate an HTML help file.

HScrollBar: Lets you create scroll bars, but are used infrequently because many controls provide the ability to display their own scroll bars.

ImageList: Allows you to display images on other controls.

Label: Displays text information that does not have a need to be edited by an end user. It's often displayed next to additional controls to label their use.

LinkLabel: Provides Web-style links for your programs.

ListBox: Contains a list of items, allowing an end user to select one or more items.

ListView: Displays a list of items with icons.

MainMenu: Provides a menu for your programs.

MonthCalendar: Displays a calendar from which dates can be picked.

NumericUpDown: Displays and sets numeric values.

OpenFileDialog: Displays a dialog box for opening files.

PageSetupDialog: Displays a dialog box for setting up pages.

Panel: Provides a panel for your program.

PictureBox: Allows you to display images in several different graphics formats, such as BMP, GIF, and JPEG, among others.

PrintDialog: Displays a dialog box for printing.

PrintDocument: Provides the ability to print a document.

PrintPreviewControl: Provides the ability to preview before printing.

PrintPreviewDialog: Displays a dialog box for print preview.

ProgressBar: Indicates progress of an action.

RadioButton: Allows the user to select from a small set of exclusive choices.

RichTextBox: Displays RichText files on a form.

SaveFileDialog: Displays a dialog box for saving files.

Splitter: Allows you to resize docked controls at runtime.

StatusBar: Displays various types of information.

TabControl: Allows you to display tabs, such as those that appear in a notebook.

TextBox: Provides a field for input or displaying text.

Timer: Provides timed functions for certain events. The Timer control is an oddity when compared to other controls in that it isn't displayed at runtime.

ToolBar: Provides a toolbar for your application.

TrackBar: Allows the user to navigate large amounts of data; also known as a "Slider" control.

TreeView: Displays a hierarchy of nodes.

VScrollBar: Lets you create scroll bars, but are used infrequently because many controls provide the ability to display their own scroll bars.

Some of the intrinsic controls are used more frequently and you are likely to become acquainted with them much faster. For example, the Button and Label control are used in almost all Visual Basic developed applications. Although some are very important, others may provide functionality that can be replaced by far superior controls.

Additional controls, known as ActiveX controls (sometimes referred to as OCX controls or OLE custom controls), provide extra functionality and can be added to the Toolbox for use in a project. These components are provided by many third-party companies or may have been provided by Visual Basic itself. Many times, these controls provide extended functionality that makes them much more powerful than the intrinsic controls. That being said, the built-in varieties offer a few advantages that cannot be overlooked. For instance, if you use a third-party control, you will need to distribute it with your application.

Form Designer

You need to have a place to assemble your controls, which is the function of forms. As you can see in Figure 2.12, the forms you work with are displayed inside the Form Designer window. When they are displayed in this way, you can place and manipulate controls.

Code Editor

Every form has a Code Editor, which is where you write the code for your program. The Code Editor can be opened in a variety of ways, such as double-clicking on a form or choosing Code from the View menu. Figure 2.13 displays a sample Code Editor.

Solution Explorer

Solution Explorer can be seen in Figure 2.14 and is provided to help you manage projects. Solution Explorer is a hierarchical, tree-branch structure that displays projects at the top of the tree. The components that make up a project, such as forms, descend from the tree. This makes navigation quick and easy because you

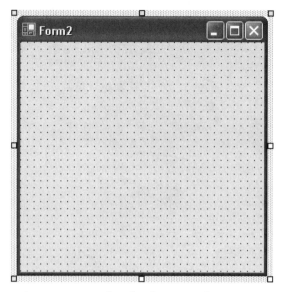

FIGURE 2.12   During development, the Form Designer displays the form you are working on.

```
        End If

        Do While ActiveRender
            dxg8.Render()
            Application.DoEvents()
            If CheckBox1.CheckState = CheckState.Checked And txtAmount.Text <> "" Then
                dxg8.Rotate(CSng(txtAmount.Text))
            End If
        Loop
    End Sub

    Private Sub CheckBox2_CheckedChanged(ByVal sender As System.Object, ByVal e As System.EventArgs) Handles CheckBox2.CheckedChanged
        dxg8.ShowFrameRate = Not dxg8.ShowFrameRate
    End Sub

    Private Sub Red_Scroll(ByVal sender As System.Object, ByVal e As System.Windows.Forms.ScrollEventArgs) Handles Red.Scroll
        dxg8.InitLights(1, Red.Value, Green.Value, Blue.Value)
    End Sub

    Private Sub Green_Scroll(ByVal sender As System.Object, ByVal e As System.Windows.Forms.ScrollEventArgs) Handles Green.Scroll
        dxg8.InitLights(1, Red.Value, Green.Value, Blue.Value)
    End Sub

    Private Sub Blue_Scroll(ByVal sender As System.Object, ByVal e As System.Windows.Forms.ScrollEventArgs) Handles Blue.Scroll
        dxg8.InitLights(1, Red.Value, Green.Value, Blue.Value)
    End Sub

    Public Sub DXZoom(ByVal percent As Single)
        dxg8.CameraPoint = dxg8.v3(dxg8.CameraPoint.x + (dxg8.CameraPoint.x - dxg8.ViewPoint.x) * percent, dxg8.CameraPoint.y + (dxg8.Cam
    End Sub

    Private Sub Button2_Click(ByVal sender As System.Object, ByVal e As System.EventArgs) Handles Button2.Click
        DXZoom(-0.1)
    End Sub

    Private Sub Button3_Click(ByVal sender As System.Object, ByVal e As System.EventArgs) Handles Button3.Click
        DXZoom(+0.1)
    End Sub

    Private Sub Form1_Closing(ByVal sender As Object, ByVal e As System.ComponentModel.CancelEventArgs) Handles MyBase.Closing
        ActiveRender = False
        Application.DoEvents()
    End Sub
End Class
```

FIGURE 2.13   Visual Basic code is written in the Code Editor.

can simply double-click on the part of the project you want to work on. For instance, if you have a project with several forms, you can simply double-click the particular form you want to view. It provides a quick and easy means of navigation.

FIGURE 2.14   You'll quickly realize the usefulness of Solution Explorer.

Solution Explorer also provides additional functions, such as the ability to add new forms by right-clicking on an open area of the Solution Explorer window, and selecting Add Form from the pop-up menu, which can be seen in Figure 2.15.

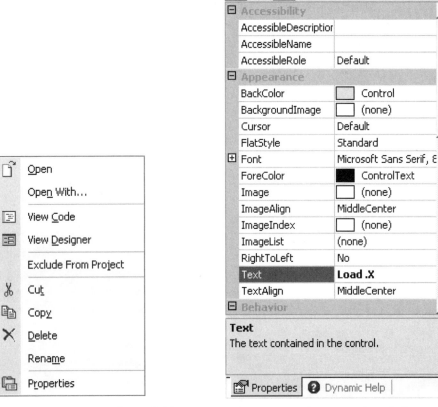

Open	
Open With...	
View Code	
View Designer	
Exclude From Project	
Cut	
Copy	
Delete	
Rename	
Properties	

FIGURE 2.15 Pop-up menus make available countless valuable features in the IDE.

FIGURE 2.16 The Properties window allows you to adjust properties for many Visual Basic objects.

Properties Window

The Properties window is used for the configuration of the controls you place on a form as well as the form itself. All of the standard Visual Basic controls have properties, and the majority of ActiveX controls do as well. As you can see in Figure 2.16, the window displays the available properties for an individual control or the forms on which they are placed. These properties can be changed as you design an application, or you can alter them in code.

SUMMARY

In this chapter, we looked at the history of Visual Basic, from its early days as a tool used mainly for prototyping, to its present standard that has made it the most popular development tool available today. We looked at the VB .NET development environment and went through some of the basic features of the IDE. In Chapter 3, Working with VB .NET, we continue our look at VB .NET by looking at its new features and we go over the basic features of the Upgrade Wizard, which walks you through the process of converting VB6 files to VB .NET. Finally, we put together the information from the first chapters to build our first project in VB .NET, the standard "Hello World!" application.

3 Working with VB .NET

In the previous chapter, we looked at the VB .NET IDE in some detail. We expand on what we learned in the previous chapter by detailing some of the differences between VB6 and Visual Basic .NET. The Windows XP Tablet PC Edition Platform Software Development Kit (SDK) can be used with VB6 or VB .NET, although this book uses .NET. This chapter helps VB6 users to get up to speed with VB .NET. First, we look at how you can use the Upgrade Wizard to move your products from VB6 to VB .NET, and then we look at the changes that have been made to the programming language.

WORKING WITH BOTH VISUAL BASIC 6.0 AND VISUAL BASIC .NET

The Visual Basic .NET and Visual Basic 6.0 IDEs can be used on the same computer and can even execute simultaneously. This is something very new as previous versions of Visual Basic caused considerable problems for one another if they were installed on the same machine. Additionally, applications written and compiled in Visual Basic .NET and Visual Basic 6.0 can be installed and executed on the same computer. Although the projects in this book have been written for Visual Basic .NET, they can be written so that they will work in VB6.

Upgrading Version 6 Projects to Visual Basic .NET

Because of the new changes associated with Visual Basic .NET, VB6 code needs to be upgraded before it can be used. Fortunately, the vast majority of the time, this is very easy because it happens automatically when you open a Visual Basic 6 project in Visual Basic .NET. This is not a perfect upgrade, however, and you'll often be left with a list of tasks that the wizard could not handle on its own.

An Upgrade Wizard, which can be seen in Figure 3.1, steps you through the upgrade process and creates a new Visual Basic .NET project. The existing Visual Basic 6 project is left unchanged. If you have Visual Basic version 5 projects, it's best to upgrade them to 6 before moving on to .NET (or version 7 as it is sometimes called).

FIGURE 3.1   The Upgrade Wizard makes it easy to convert version 6 projects to Visual Basic .NET.

When your project is upgraded, the language is modified for any syntax changes and your Visual Basic 6.0 forms are converted to Windows Forms. Depending on your application, you may need to make minor changes to your code after it is upgraded. Many times, this can be necessary because certain features either are not available in Visual Basic .NET or the features have changed significantly enough to warrant manual changes.

After your project is upgraded, Visual Basic .NET provides an "upgrade report" to help you make changes and review the status of your project. The items are displayed as tasks in the new Task List window, so you can easily see what changes are required, and navigate to the code statement simply by double-clicking the task. Many times, the document recommendations simply represent good programming practices, but they also identify the Visual Basic 6 objects and methods that are no longer supported.

PROBLEMS WITH CODE UPGRADES

When your VB6 code is upgraded to Visual Basic .NET, it follows a specific set of rules. The following rules list some basic information that you should keep in mind if you are currently planning to upgrade a VB6 project to .NET.

Variant to Object

Previous versions of Visual Basic supported the Variant data type, which could be assigned to any primitive type. In fact, it is the default data type if a variable wasn't declared in VB6. VB .NET converts variants to the Object data type.

Integer to Short

In Visual Basic .NET, the data type for 16-bit whole numbers is now Short. The data type for 32-bit whole numbers is now Integer, and Long is now 64 bits.

Table 3.1 shows a few examples.

TABLE 3.1   VB6 to .NET data type conversions

VB6	VB .NET
Dim A as Integer	Dim A as Short
Dim B as Long	Dim B as Integer
N/A	Dim A as Long
Variant	N/A (Use new 'Object')
Currency	N/A (Use Decimal or Long)
N/A	Decimal
String	String (doesn't support fixed-length strings)

APIs

The vast majority of API calls expect 32-bit values if they take numeric arguments. With the previous section in mind, you can see that problems are sure to arise. For instance, in VB6, a 32-bit value is a Long data type, whereas in .NET, a Long is 64 bits. You'll have to use Integer as the data type in .NET to make the calls correctly. According to Microsoft documentation, many APIs will no longer be callable from VB or may have replacements.

The Upgrade Wizard tries to correct API calls by creating wrappers for them. This is not a good idea and you should look at every API call individually to make any changes you need.

The following code samples are an example of an API call under VB6 and under VB .NET.

Listing 3.1   Visual Basic 6.

```
Private Declare Function GetVersion Lib "kernel32" () As Long
Function GetVer()
    Dim Ver As Long
    Ver = GetVersion()
    MsgBox ("System Version is " & Ver)
End Function
```

Listing 3.2   Visual Basic .NET.

```
Private Declare Function GetVersion Lib "kernel32" () As Integer
Function GetVer()
Dim Ver As Integer
Ver = GetVersion()
MsgBox("System Version is " & Ver)
End Function
```

Newly Introduced Keywords

VB .NET introduces several new keywords that have no counterpart in VB6. Table 3.2 details some of them:

TABLE 3.2   New keywords in Visual Basic .NET

Keyword	Notes
Catch	New error handling; indicates code to use to process errors
Char	New character data type
Finally	New error handling; indicates code to use to run regardless of errors
Imports	Makes an object hierarchy (namespace) available in a module
Inherits	Points to a base class for inheritance

TABLE 3.2   New keywords in Visual Basic .NET (*continued*)

Keyword	Notes
MustOverride	Indicates that any class that derives from this class must supply an override for this member
MyBase	References the base class for use by subclass code
Namespace	Specifies a namespace for a module
Overloads	Indicates there's more than one version of a function and the compiler can distinguish among them by the input parameters
Overrides	Indicates a member overrides the identically named member in the base class
Overridable	Allows a member to be overridden in any class derived from the base class
Protected	Is only available to classes derived from this class
ReadOnly	Is used in a property that contains only a "Get"
Shared	Indicates all instances of a class should share a variable in a class
Throw	New error handling; allows you to raise an error
Try	New error handling; starts code with error handling enabled
Webmethod	Tags a method as part of a publicly available Web service
WriteOnly	Is used in a property that contains only a "Set"

Removed from .NET

Like the additions from the previous section, there are also some items that have been removed in VB .NET. The following list details some of the removed keywords and statements:

- VarPtr
- ObjPtr
- StrPtr
- LSet
- GoSub
- Let
- Is Missing

- DefBool
- DefByte
- DefLng
- DefCur
- DefSng
- DefDbl
- DefDec
- DefDate
- DefStr
- DefObj
- DefVar
- On x Goto

Table 3.3 details some commands that have equivalents in VB .NET:

TABLE 3.3 Equivalent commands in VB6 and VB .NET

VB6	*VB .NET Namespace*	*Method/Property*
Circle	System.Drawing.Graphics	DrawEllipse
Line	System.Drawing.Graphics	DrawLine
Atn	System.Math	Atan
Sgn	System.Math	Sign
Sqr	System.Math	Sqrt
Lset	System.String	PadRight
Rset	System.String	PadLeft
Rnd	Microsoft.VisualBasic.Compatibility.VB6	Rnd
Round	Microsoft.VisualBasic.Compatibility.VB6	Round
DoEvents	System.Winform.Application	DoEvents

This is certainly not an exhaustive list as there are sure to be other items that have been added and removed. However, this list should serve as a good focal point for your work. There are a couple of additional commands that are no longer supported. One in particular is the Set command, which has been removed so that the following code in VB 6

```
Set objObject = objAnotherObject
```

will be changed to:

```
objObject = objAnotherObject
```

Another widely used command that has been changed is the Debug command, which was used as follows in VB6:

```
Debug.Print
```

In VB .NET, this is now:

```
Debug.Write
Debug.WriteLine
```

Arrays

The use of arrays in VB .NET has also changed. In Visual Basic 6, if you declare an array like the following, you would get 11 items from 0 to 10:

```
Dim number(10) as Integer
```

Within VB .NET, the same array only gives you 10 items from 0 to 9.

Default Properties

In Visual Basic 6, a control or object had a default property that wouldn't need to be specified. For instance, if you wanted to set a TextBox equal to a string, you would simply use:

```
txtInformation = "This is a string"
```

Visual Basic .NET doesn't support default properties. So instead, you have to make sure to specify it as follows:

```
txtInformation.Text = "This is a string"
```

References to Form Controls

Controls in Visual Basic 6 were public. That is, you could simply reference a control on form 1 inside the Code Editor on form 2 by simply using form1. textbox1.text. In VB .NET, you'll have to create a public Let and Get property procedure for every control property you want to have access to.

Get and Let are now combined in VB .NET, so instead of being two separate property procedures, you create one.

Listing 3.3   Visual Basic 6 code.

```
Property Get PropertyA() As Integer
    m_PropertyA = PropertyA
End Property

Property Let PropertyA(NewValue As Integer)
    m_PropertyA = NewValue
End Property
```

Listing 3.4   Visual Basic .NET code.

```
Property PropertyA() As Short
  Get
      m_PropertyA = MyPropertyA
  End Get
  Set
      m_PropertyA = Value
  End Set
End Property
```

Forms and Controls

Visual Basic .NET forms are now called Windows Forms. The following list gives you an idea of the differences between version 6 and .NET forms:

- Windows Forms do not support the OLE container control.
- Windows Forms only support true-type and open-type fonts.
- There are no shape controls in Windows Forms. Shape controls have been upgraded to labels.
- Drag-and-drop properties of VB6 do not work on Windows Forms.
- Windows Forms have no support for Dynamic Data Exchange (DDE).
- There is no line control in Windows Forms.
- Windows Forms do not have access to form methods, such as Circle or Line.
- Windows Forms do not support the PrintForm method.
- Clipboards are different and cannot be upgraded from version 6 to .NET.

SUMMARY

In this chapter, we looked at some of the changes that have been made to VB .NET. There are complete books written on this subject, so this chapter should not be considered an all-inclusive list of changes. However, it should serve its purpose of providing you with the ideas that have been altered and will help you in subsequent chapters. In Chapter 4, Basics of the .NET Framework, we spend a little time looking at the Visual Basic .NET Framework. After we have a little more background information, we move on to writing several complete applications in VB .NET.

4 Basics of the .NET Framework

Using the .NET Framework, Microsoft Visual Basic developers can build robust applications that were very difficult to write in previous versions of Visual Basic. The .NET Framework is Microsoft's latest offering in the world of cross-platform development. In doing so, there have been many changes to the VB language and even the philosophy of developing applications in VB .NET has been forever altered. Although the changes are very positive for VB programmers, they can be a little difficult to become accustomed to.

The .NET Framework is essentially a combination of the common language runtime (CLR) and the standard classes that are available to .NET programmers. The CLR is the execution environment for all programs in the .NET Framework. It is similar in nature to the Java™ virtual machine (VM). The Runtime classes provide hundreds of prewritten services that clients can use as the basic building blocks for an application. We look at these key concepts in a little more detail in the following sections.

COMMON LANGUAGE RUNTIME

The CLR is the new runtime environment shared by every .NET application. Its purpose is to manage code execution along with all of the services that .NET provides.

The CLR is very similar to the concepts used in other languages, such as the VB6 runtime library, a Java virtual machine, or even the Microsoft Foundation Class (MFC) library for C++.

One of the nice features of .NET is that there are many languages that can take advantage of this same runtime (currently C# ["C-Sharp"], VB, and C++). It is the purpose of this runtime to execute the code for any language that was developed for it. Code that was developed for this is called managed code, which can be simply thought of as a relationship between the language and the runtime itself. Because all

.NET languages use this same runtime, there is no longer a need for language-specific runtimes. Further enhancing this is the ability of C# programmers and VB programmers to share their code with no extra work.

With the shared runtime, there is a common file format for .NET executable code, which is called the Microsoft intermediate language (MSIL, or just IL). IL is a semicompiled language that is compiled into native code by .NET when the program is executed. This is similar to what all versions of Visual Basic did prior to version 5. This previous method for compiling was known as pseudocode or p-code. With the CLR, the best features of pseudocode and compiled languages are available.

The following list details some of the benefits of using the CLR:

Consistent programming model: All application services are offered via a common object-oriented programming model. This differs from past programming models in which some functions are accessed via DLLs, whereas others are accessed via the COM object.

Simplified programming model: By removing archaic requirements of the Win32 API, .NET seeks to greatly simplify the development process. More specifically, a developer is no longer required to delve into GUIDs, HRESULTS, and so on.

DLL problems: When programmers are using DLLs, they are always concerned with installing components for a new application that can overwrite components of an old application. The old application often exhibits strange behavior or stops functioning altogether because the DLL has changed the way it works. The .NET architecture removes this common problem, and if the application runs after its installation, then the application should always run without the headaches caused by DLL versions.

Platforms: Today, there are many different flavors of Windows, including the three Windows XX variations (Windows® 95, Windows® 98, and Windows® Me), Windows NT® 4.0, Windows® 2000, Windows XP, and Windows® CE. Most of these systems run on x86 CPUs, with the exception of Windows CE devices, which work with Million Instructions Per Second (MIPS) and StrongARM processors. A .NET application that consists of managed code can execute on any platform that supports the .NET common language runtime. This opens some interesting possibilities for the Windows platform and could further be enhanced if the .NET platform is ever ported to platforms such as Linux or the Mac OS.

Reuse of code/language integration: .NET allows languages to be integrated with one another. For example, it is now entirely possible to create a class in

C++ that derives from a class first implemented in Visual Basic. The Microsoft Common Language Specification ensures that all .NET languages are implemented in a manner that allows this. Although there are currently the three .NET languages (C#, VB, and C++), there are sure to be many more available in the near future with companies other than Microsoft producing compilers that also target the .NET common language runtime.

Resource management: A common error among applications occurs when a developer forgets to free up resources once they are no longer needed. This can cause a great deal of inconsistency in an application and is sometimes one of the most difficult bugs to track down. Again, the .NET CLR comes to our rescue by automatically tracking resource usage.

Types: The .NET common language runtime can verify that all your code is type safe, which simply means that allocated objects are always accessed in compatible ways. This actually helps in eliminating many common programming errors, but is also advantageous because of the protection of exploitation of buffer overruns used by many hackers.

Error handling: One of the most difficult aspects of programming in Windows is the various ways in which error messages can be reported. For example, some return an HRESULT, whereas others may return a Win32 error code. In .NET, all errors are reported via exceptions, which greatly simplifies maintaining code.

Deployment: This could go side by side with the DLL problems mentioned previously. Past Windows-based applications were incredibly difficult to install and deploy. There can be many files, including data, several DLL files, and Registry settings to deal with—to name a few of the more common examples. .NET components are no longer referenced in the Registry and installing most .NET components can be as simple as copying the files to a directory. Uninstalling can sometimes be an even more difficult task in past Windows-based applications, but with .NET, you can simply delete those same files you installed.

Common type system: The CLR greatly simplifies cross-language communication through the introduction of the common type system (CTS). The CTS defines all of the basic types that can be used in the .NET Framework and the operations that can be performed on those types. Although an application can create a more complex type, it must be built from one of the CTS defined types, which are classes that derive from a base class called `System.Object`. We look at the CTS in more detail in Chapter 5, Introduction to the VB .NET Language.

THE FOUNDATION FOR .NET DEVELOPMENT

As was previously mentioned, you can think of the .NET Framework as a foundation on which you build and run applications. Having such a foundation makes it easier to build applications while using a consistent, simplified programming model. In earlier versions of VB, we spent a great deal of time using the Win32 API to do things that we could not natively access in the standard VB functions—things such as accessing Registry keys or creating irregularly shaped forms. Another option was third-party ActiveX controls that are actually very easy to use, but are often expensive for a single project.

This flexibility offered in the Win32 API was a blessing for VB programmers, allowing access to a variety of functions that were otherwise unavailable. Unfortunately, with this power also came a great deal of problems and confusion. Learning the API can be difficult because the calls are complex and, many times, the available examples were written for C++. This made calling the API very error-prone, often resulting in crashed computers and lost hours of coding.

Another problem was the deployment of these applications. After you finally managed to get the API calls working, you then had to deploy the appropriate DLL files with the correct version numbers. You can quickly see why this common framework makes sense from a development standpoint. You still have access to most of the powerful features of the API, without actually knowing how to implement them. In addition, deployment is now very simple because you don't need to concern yourself with how to handle the various DLL files.

API TO VB .NET

Although the API has not become obsolete, you need to have a quick way to access the various classes that implement the functions you have used in the past. The following BitBlt API call details a portion of a common .NET equivalent:

BitBlt performs a bit-block transfer of the color data corresponding to a rectangle of pixels from the specified source device context into a destination device context. Here is a sample API call and a .NET equivalent:

Listing 4.1   WinAPI.

```
Declare Function BitBlt Lib "gdi32" Alias "BitBlt" ( _
        ByVal hDestDC As Long, _
        ByVal x As Long, _
```

```
            ByVal y As Long, _
            ByVal nWidth As Long, _
            ByVal nHeight As Long, _
            ByVal hSrcDC As Long, _
            ByVal xSrc As Long, _
            ByVal ySrc As Long, _
            ByVal dwRop As Long _
    ) As Long
```

Listing 4.2   .NET.
_____

```
    System.Drawing.Graphics.DrawImage
```


SUMMARY

In this chapter, we looked at some of the basics of the CLR, CST, and the overall .NET Framework, a topic that encompasses so much information that we could devote an entire volume to its study. In Chapter 5, Introduction to the VB .NET Language, we look at some of the syntax of the VB .NET programming language and take an even greater look at the differences between VB6 and .NET.

5 | Introduction to the VB .NET Language

Although this book is focused on the creation of applications, it's important that you have some fundamental knowledge of the VB .NET language. This chapter introduces you to some of the basic principles. For those of you with programming experience, you can probably skim over this chapter and move on. For those without any programming experience, this chapter will help you with some of the basic concepts used in most VB applications.

VARIABLES

Variables are used in almost every VB application. They are simply used to store data in a memory location that you can access and manipulate as needed. For example, suppose you are developing an application that adds two numbers together. When writing the code, you could temporarily store the values of the two numbers in separate variables with a third variable holding the resultant value. Instead of referring to an actual memory location, VB allows us to use a variable name that we can declare to refer to these values. Declaring a variable is as simple as creating a name along with a specific data type. When you declare a variable, Visual Basic allocates a certain amount of memory for the variable to store the data. It is the data type that determines exactly how much memory is being put aside.

 If you don't specifically declare a data type, VB .NET assigns it to the Object *data type. This is not a suggestion to do so; as for readability, you should always declare the data type, even if it is of type* Object.

Table 5.1 details some of the variable types, the range of values they can store, and the memory that is allocated.

TABLE 5.1   Data types, size, and range of memory

Type	Size	Range
Boolean	4 bytes	True or False
Byte	1 byte	0–255 unsigned
Char	2 bytes	0–65,535 unsigned
Date	8 bytes	1/1/1 CE to 12/31/9999
Decimal	12 bytes	+/– 79,228,162,514,264,337,593,543,950,335 with no decimal point; +/– 7.9228162514264337593543950335 with 28 places to the right of the decimal; smallest nonzero number is +/– 0.0000000000000000000000000001
Double	8 bytes	–1.79769313486231E308 to –4.94065645841247E-324 for negative values; 4.94065645841247E-324 to 1.79769313486232E308 for positive values
Integer	4 bytes	–2,147,483,648 to 2,147,483,647
Long	8 bytes	–9,223,372,036,854,775,808 to 9,223,372,036,854,775,807
Object	4 bytes	Any object type
Short	2 bytes	–32,768 to 32,767
Single	4 bytes	–3.402823E38 to –1.401298E-45 for negative values; 1.401298E-45 to 3.402823E38 for positive values
String	10 bytes	+ 0 to approximately 2 billion Unicode characters (characters in string * 2)

With some of this basic information out of the way, we'll look at how variables are actually declared using the Dim keyword.

As in previous versions of Visual Basic, you use the Dim keyword. The following are some common examples of declaring variables:

```
Dim x as Single
Dim txt as String
Dim str as string
Dim oObj as Object
```

By default, when you declare a variable in VB, it is initialized to a standard value (numeric variables are set to 0, strings are initialized to an empty string " ", and object variables are initialized to nothing). In VB .NET, you can now initialize variables to something other than their defaults when you declare them. Here are a few examples of initializing variables when declaring them:

```
Dim x as Single = 1.5
Dim txt as String = "Bob"
Dim Answer as Boolean = "True"
```

Variable declarations are usually very simple, although they can get a little more complicated when you invoke an object's constructors. Different constructors use different arguments to initialize the object. For example, suppose you need to initialize a string with 50 asterisks ("*"). You could manually type in 50 of them and it would work just fine, although you could also use a String variable constructor as follows:

```
Dim txt3 As New String("*", 50)
```

You can see that this is much easier than attempting to type out 50 asterisks and is also much more readable.

When multiple variables are declared on the same line, then its type is the same as the rest of the variables on the line. For example:

```
Dim x,y,z as Integer
```

This gives x, y, and z the Integer data type. You can take this a step further as well:

```
Dim x,y,z as Integer, a,b as String
```

This sets the x, y, z types to Integer and a, b types to String. In earlier versions of VB, these types of declarations could have caused some problems, so this is a welcome addition.

CTS

Before moving on, we look at a more exhaustive list of the data types supported in the .NET CTS, as shown in Table 5.2. It's worth noting that the CTS data types are either structures (which are value types) or classes (which are reference types).

TABLE 5.2 Data types supported in the .NET CTS

Data Type	CTS Type	Type	Storage	Value Range
Boolean	System.Boolean	Value (Structure)	2 bytes	True or False
Byte	System.Byte	Value (Structure)	1 byte	0 to 255 (unsigned)
Char	System.Char	Value (Structure)	2 bytes	A character code from 0 to 65,535 (unsigned)
Date	System.DateTime	Value (Structure)	8 bytes	January 1, 1 CE to December 31, 9999
Decimal	System.Decimal	Value (Structure)	12 bytes	+/-79,228,162,514,264,337,593,543,950,335 with no decimal point; +/- 7.9228162514 6433759354950335 with 28 places to the right of the decimal; smallest nonzero number is +/-0.0000000000000000000000000153
Double (double-precision floating point)	System.Double	Value (Structure)	8 bytes	-1.79769313486231E308 to -4.940656458 41247E-324 for negative values; 4.94065645 841247E-324 to 1.79769313486232E308 for positive values
Integer	System.Int32	Value (Structure)	4 bytes	-2,147,483,648 to 2,147,483,647
Long (long integer)	System.Int64	Value (Structure)	8 bytes	-9,223,372,036,854,775,808 to 9,223,372,036,854,775,807
Object	System.Object	Reference (Class)	4 bytes	Any type can be stored in an Object variable
Short	System.Int16	Value (Structure)	2 bytes	-32,768 to 32,767
Single (single-precision floating point)	System.Single	Value (Structure)	4 bytes	-3.402823E38 to -1.401298E-45 for negative values; 1.401298E-45 to 3.402823E38 for positive values
String (variable-length)	System.String	Reference (Class)	10 bytes + (2 * string length)	0 to approximately 2 billion Unicode characters
User-Defined Type (structure)	(inherits from System.Value Type)	Value (Structure)	Sum of the sizes of its members	Each structure member has a range determined by its data type and is independent of the ranges of the other members

SCOPE

The scope of a variable determines where in a program it is visible to the code. Variables and constants (we look at constants in the next section) both have a scope, which allows a programmer to decide when a variable can be referred to in the rest of the program.

Block-Level and Procedure-Level Scope

If a variable is declared inside a block of code, then the variable has block-level scope. This basically means that the variable is visible only within that block of code. A block of code, in this instance, refers to a set of programs that is terminated by a loop. Look at the following example that would give an error if it was executed:

```
If x  0 Then
      Dim intAmount As Integer
      intAmount = 50
End If
MsgBox CStr(intAmount)
```

This would give an error because the value of intAmount cannot be seen outside of the block of code encompassed by the If...End If loop.

If you declare a variable inside a procedure but not within the constraints of a loop, the variable is said to have procedure-level scope. This allows us to utilize the variable within the procedure, but once you get outside of it, the variable is again invisible to the rest of the program. A nice feature of procedure-level variables is that you don't really have to worry as much about the naming of them because each procedure can name their variables exactly the same. Because you cannot see them outside of the procedure, the code does not cause a problem.

Module-Level and Project-Level Scope

Module-level and project-level declarations can get a little more confusing because the modules themselves are declared using one of several access modifiers (Private, Public, or Friend). Don't concern yourself with these ideas at this time, but remember that if you declare a module as a Friend, and then declare a variable as a Public inside the module, the variable takes on the attributes of the module in which it was declared and, thus, has a Friend scope.

A variable that is declared in the declarations section of a standard module using the Private access modifier has module-level scope. It is visible in the entire module, but when you are outside of the module, it is invisible to the rest of the program code. Now, if you were to create the same variable by using the Friend access modifier, the variable would be visible in the entire project (project-level scope). It is not visible to other projects, however. A third possibility exists if you declare the same variable as a Public modifier (the module would also need to be Public); the variable is visible to this project and any additional projects that reference this one.

Lifetime

Many people confuse a variable lifetime with its scope, but the differences are actually very clear. The lifetime of a variable refers to what time during program execution a variable is valid. As you know, scope refers to where the variable can be seen by other code. A static variable is a special type of variable, which has a lifetime that exists during the entire program execution. Previous versions of VB lacked this feature, although you could implement something similar using a workaround.

CONSTANTS

Constants are similar to variables, although the value that is assigned to a constant does not get changed during program execution. You can use constants instead of hard coding the values directly into your code as it's much easier to make changes to your code if you approach it in this manner.

For example, suppose you are developing an application that uses the value of pi (3.1415) for calculations. Instead of adding the value 3.1415 to every line that uses it for a calculation, you can store it in a variable. Look at the following code example:

```
Const X as Double = 3.1415
Dim Y as Integer
Dim Answer as Double

For Y = 1 to 100
    Answer = Const * Y
Next Y
```

You could have replaced Const with the actual value 3.1415 in the Answer = line. This would be very easy for such a simple example. However, suppose you use this value in 40 or 50 calculations in different modules and procedures in your code. Now, assume that you have been asked to shorten 3.1415 to 3.14 for the calculations. You could change this in a single location in your program by changing the X constant, or you could go through hundreds of lines of code searching for 3.1415. Although you could eventually change all of them, it saves a great deal of time and is much more reliable to use a constant.

STRUCTURES

Now that we have looked at declaring variables using the built-in data types, we're going to take a moment to look at the ability to create your own custom data types using a structure. A structure contains members that can be any data type. The members each have their own names, allowing you to reference them individually. The following example creates a structure called Customer:

```
Structure Customer
Dim ID As Long
Dim Name As String
Dim Address As String
Dim City As String
Dim State As String
End Structure
```

After you have created a structure, you can use it within your program. The following example gives you an idea on how to use these structures:

```
Dim cust as Customer

Cust.ID = 1000
Cust.Name = "Clayton Crooks"
Cust.Address = "12345 South Main Street"
```

These lines created a variable called cust as type Customer. We then had access to the individual members and assigned them values of "Clayton Crooks", 1000, and "12345 South Main Street".

CONVERTING BETWEEN DATA TYPES

In a project, you'll often be faced with the prospect of converting data from one format to another. For example, suppose you have an integer value that you want to store in a text box. You'll need to convert the data from an Integer to a String. The process of converting the values is known as casting, or simply conversion. A cast can be one of two distinct types: widening or narrowing. A widening cast is one that converts to a data type that can handle all of the existing data. That is, if you convert from Short to Integer, the Integer data type can handle everything that was stored in the Short data type. Therefore, no data is lost. The narrowing cast is obviously the opposite of this, in which data is lost in the conversion process.

With VB .NET, data conversions are made in one of two ways: implicitly or explicitly. An implicit conversion occurs without interaction from the developer. For example, look at the following code:

```
Dim X as Long
X = 99
```

Although the value 99 that is stored in X is obviously an Integer, it is implicitly converted to a Long for storage in the variable. VB .NET did this on its own without any interaction. Explicit conversion requires calling one of the many VB .NET conversion functions. Table 5.3 details these functions:

TABLE 5.3   VB .NET conversion functions and descriptions

Function	Example	Description
CBool	newValue = CBool(oldValue)	Converts any valid String or numeric expression to Boolean. If a numeric value is converted, it results in True if the value is nonzero or False if the value is zero.
CByte	newValue = CByte(oldValue)	Converts a numeric data type from 0 to 255 to a byte. Any fractions are rounded.
CChar	newValue = CChar(oldValue)	Returns the first character of a string that is passed to it.
CCur	newValue = CCur(oldValue)	Converts a data type to Currency.
CDate	newValue = CDate(oldValue)	Converts a data type to Date or Time.
CDbl	newValue = CDbl(oldValue)	Converts a data type to Double.
CInt	newValue = CInt(oldValue)	Converts a data type to Integer while rounding fractional portions.
CShort	newValue = CShort(oldValue)	Rounds any fractional part while converting to a Short.
CSng	newValue = CSng(oldValue)	Converts a data type to Single.
CStr	newValue = CStr(oldValue)	Converts a data type to String.

ARRAYS

When you develop applications, you're often faced with the need to store multiple instances of like data. An array is a set of variables that are represented by a single name. The individual variables are called elements and are identified from one another via an index number.

Arrays have a lower bound and an upper bound and have changed slightly in Visual Basic .NET. This is one of those areas in which VB6 programmers need to be careful. The lower bound for an array is always 0, and unlike previous versions of VB, you cannot change the lower bound. For example, suppose you have an array of 10 elements. This suggests that the lower bound is 0 and the upper bound is 9.

Declaring an array is similar to declaring other variables:

```
Dim Name(50) As String
```

This creates an array of 50 elements that are each a string. The lower bound is 0 and the upper bound is 49.

Like variables, you can also initialize the array when you declare it. The following code declares an array of integers and initializes the values.

```
Dim Num1(5) As Integer = {9,8,7,6,5}
```

Alternatively, you could also declare this as follows:

```
Dim Num1() As Integer = {9,8,7,6,5}
```

This dimensions the variable to the appropriate number of elements automatically.

You can access the individual elements of an array as you would other variables. For instance, to set the third element to 0:

```
Num1(3) = 0
```

Multidimensional Arrays

An array isn't limited to a single dimension. They can have multiple dimensions that allow you to create a grid. VB .NET can have up to 60 dimensions, although this is probably unrealistic for any real-world use. The following example gives you an idea of dimensioning an array:

```
Dim grid(3, 3) As Integer
```

To access the information in this array to retrieve or assign values, you use the combination of numbers to identify the element:

```
grid(0, 0) = 1
grid(1, 1) = 0
```

You can also declare an array with uneven elements. For example, if you want to declare an array that holds the following data:

1	2	3	4	5
1	2	3	4	5
1	2	3	4	5

You would use the following code:

```
Dim arr(3,5) As Integer
```

To initialize a multidimensional array when you declare it, you leave the parentheses empty with the exception of a comma for each additional array dimension. For example, to initialize an array, you could use something like this:

```
Dim arr(,) As String = {{"1", "1", "1"}, {"2", "2", "2"}}
```

Dynamic Arrays

In VB .NET, all arrays are dynamic. The declared size is only the initial size of the array, which can be changed as needed. (Please note that you cannot change the dimensions of an array after declaration.) To change the size of an array, you use the ReDim statement. The process of changing the size of an array is known as redimensioning. This is confusing because we've already mentioned that you cannot actually redimension an array.

 There are two functions that can help when redimensioning arrays: UBound *and* LBound. UBound *returns the upper limit, whereas* LBound *returns the lower bound.*

It is actually very simple to redimension an array. Let's start with an array dimensioned as follows:

```
Dim Nums(10, 10) As Integer
```

To redimension this, you can use the following:

```
ReDim Nums (100,100) as Integer
```

When you redimension an array, all of the data in it is lost. If you need the data, you can use the Preserve keyword to keep the existing data. If you use the Preserve keyword, you can only change the last coordinate:

This is OK:

```
ReDim Preserve Nums (10,100) as Integer
```

Whereas this is not:

```
ReDim Preserve Nums (100,100) as Integer
```

In VB6, you could have an array of controls that were very similar to the arrays we've been looking at. The new changes to the event model make control arrays a thing of the past as the event model in Visual Basic .NET allows any event handler to handle events from multiple controls. For example, if you have multiple Button controls on a form (called Button1 and Button2), you could handle the click events for both of them as follows:

```
Private Sub MixedControls_Click(ByVal sender As System.Object, ByVal e
As System.EventArgs) Handles Button1.Click, Button2.Click
```

LOOPS

When you develop applications in VB .NET, you need a way to manage the execution of the programs. For example, you might need to execute code a certain number of times, or you may need to check a condition to see if you should be executing the code. For both cases, we use loops.

If...Then...Else

If . . . Then . . . Else loops allow you to test a condition before executing a block of code. For the block of code to execute, the condition must evaluate to True. Let's take a look at an example of an If...Then statement:

```
Dim X as Integer

X = 0
If X < 1 Then
    X = 6
End If
```

This simply checks the value of x and if it is less than the value of 1, it assigns the value 6 to it. The program code execution then continues with the next step.

Another example allows us to look at the If...Then...Else statement:

```
Dim Cost1 As Integer
Dim Cost2 As Integer
Dim BuyIt As Boolean

Cost1 = 50
Cost2 = 75

If Cost1 < Cost2 Then
      BuyIt = False
Else If Cost2>=Cost1 Then
      BuyIt = True
End If
```

The previous code is a very simplistic example, but serves its purpose to show how the ElseIf statement can be added to check multiple conditions. For If...Then statements, you need to use comparison operators to check the various conditions. In the previous example, we use the less than (<), greater than (>), and equal to (=) operators. The following list details the various types of comparison operators:

- =   Equal to
- <   Less than
- <=   Less than or equal to
- >   Greater than
- >=   Greater than or equal to
- <>   Not equal to

There are two additional operators that are worth mentioning: the Is and Like operators. The Like operator is used to compare a string to a string pattern rather than an exact copy of the string, whereas the Is operator is used to compare if two object references point to the same object. When you use the Like operator, you can use wildcards for pattern matching, as shown in Table 5.4.

The Like operator gives you many options when looking for patterns in a string. By testing the result of the condition, you can return a value of True if it is found; otherwise, a value of False is returned. Table 5.5 shows a few examples.

Sometimes, a single expression in an If...Then...Else statement is not enough. You can use multiple expressions to create a single True or False value for an If...Then...Else statement. You can use the logical operators to create compound expressions that, as a whole, return a single Boolean value. Table 5.6 lists the logical operators.

TABLE 5.4   Wildcard character pattern matches

Character	Function
?	Matches a single character
*	Matches all or none characters
#	Matches a single digit
[character list]	Matches any single character in the character list
[! Character list]	Matches any single character NOT in the character list

TABLE 5.5   Like operator examples

Value	Operator	Condition	Result
"pqrs"	Like	"p*p"	False
"pqrs"	Like	"p*s"	True
"pqr"	Like	"p?r"	False
"pqr"	Like	"p?r"	True
"pqr"	Like	"p#r"	False
"pqr"	Like	"p#r"	True
"aQa"	Like	"a[a-z]a"	False
"aba"	Like	"a[a-z]a"	True
"aba"	Like	"a[!a-z]a"	False
"aBa"	Like	"a[!a-z]a"	True

TABLE 5.6   Logical operators

Logical Operator	Result
And	Both expressions should be True to get a True result
Not	Expression must evaluate to False for a True result
Or	Either of the expressions must be True for a True result
Xor	One expression can be True for a True result

You may need to test multiple conditions in an `If...Then` statement. VB .NET provides this functionality as well. Table 5.6 details the available operators.

You can combine the logical operators to create compound expressions. Table 5.7 takes a look at some examples of compound expressions.

TABLE 5.7 Combining logical operators to create compound expressions

Operators	*Result*
0=0 And 1=2	False
0=0 And 1<2	True
0<0 Or 1=2	False
0=0 Or 1=2	True
(0=0 And 1=2) Or 1=3	False
(0=0 And 1=2) Or 1<3	True
Not 1=1	False
0=0 And Not 1=2	True
0=0 Xor 1<2	False
0=0 Xor 1<2	True

Select Case

Another type of loop that is very useful is the `Select Case` statement, which is very similar in functionality to the `If...Then` statement. The `Select Case` is most often used as a replacement for an `If...Then` statement that gets too long. As an `If...Then` statement grows in length, it gets much more difficult to read. The `Select Case` statement is much easier to read as you'll see in the following example:

```
Select Case X
      X<= 5
      Y = 10
      X>5 And X<=10
      Y = 15
      X>10 And X<=15
      Y = 20
      X>15 And X<=20
      Y = 25
End Select
```

For **Loops**

The For loop is used when you want to execute a block of code a finite number of times. For loops are often used when you want to read or write to or from an array. For loops use a counter to keep track of the number of iterations. A Start value is the beginning value of the counter and an End value is used as the maximum value for the loop. There is also an optional Step value (default is 1) that instructs the loop how much to increment the counter during each pass of the loop. The Step value can be a negative or positive number. For example, here is a simple loop:

```
Dim I As Integer

For I = 1 to 100
    Q = I * I
Next I
```

 Another variation of the For *loop is the* For...Each *loop. It is most often used with arrays to loop through every item in the array.*

While **Loops and** Do Until **Loops**

For loops work well if you know the exact number of times you want to execute a loop beforehand. There is another loop, called the While loop, that allows you to execute code blocks without knowing the values. It executes the loop until the condition that is being tested remains True.

This loop is started with the Do While keywords. A condition is checked to see if it is True prior to each execution of the loop. A variation of the While loop is the Do Until statement, which executes until the condition is True. A third variety, known as While...Wend, is no longer available in Visual Basic .NET. Here is a quick example of a Do While Loop in VB .NET:

```
Dim X As Integer = 0

Do While X < 1000
    X = X + 100
Loop
```

This loop executes until the x value is greater than or equal to 1000 (it actually checks to see if x < 1000 and, if so, it continues). You don't always have to wait until the loop is finished before you exit it. You can use the Exit Do statement at any time. Here is an example:

```
Dim X As Integer = 5
Dim Y As Integer = 10

Do While X < 10
X = X+1
Y = Y + 1

        If Y > 12 Then
        Exit Do
        End If
Loop
```

BASICS OF FUNCTIONS AND PROCEDURES

Functions and procedures are blocks of code that can be called at any time in an application. They typically perform a set type of function that is useful to the application. For example, you could have a function that calculates the square root of a number. You would then call this function whenever you needed to perform this calculation. We use functions and procedure throughout the book, so we'll go over a few simple examples now:

```
Function LessThan500(ByVal num as Integer) As Boolean
        If num<500 Then
        Return True
        Else
        Return False
        End If
End Function
```

This function returns a Boolean value of True or False depending on if the value of the number that is passed to the function is less than 500. To call this function in your own program, you can do so like any of the built-in varieties. For example, you could use a variable to pass it like this:

```
Dim LT500 As Boolean
Dim X As Integer

X = 10
LT500 = LessThan500(X)
```

This passes the value of 10, and because it is less than 500, it assigns True to the value of LT500.

Another way to return a value from a function is to assign the function name itself to a value. Using the same previous example, notice the lack of the word "Return":

```
Function LessThan500(ByVal num as Integer) As Boolean
        If num<500 Then
        LessThan500 = True
        Else
        LessThan500 False
        End If
End Function
```

You can use the method that is easier for you.

There is a big difference in how parameters are passed to functions from the earlier versions of VB. In the previous versions of Visual Basic, parameters were passed by reference by default. In Visual Basic .NET, parameters are passed by value by default.

At first, this might not seem like a big difference, but it can be in your VB .NET applications. If a parameter is passed by value, any changes made to the value occur only within the function and the rest of the code in the application doesn't see them. On the other hand, if a parameter is passed by reference, if changes are made to the value, the change is seen by the rest of your program.

Sub procedures are similar to functions but they do not return any values. You call them exactly the same as you would a function, such as:

```
somproc(10)
```

You can create a simple Sub procedure just as you would a function:

```
Sub somproc(ByVal x as Integer)
If x > 5 then
    Temp = 5
Else
    Temp = 10
End Sub
```

You'll notice that it looks almost identical to the previous example of a function with the exception of returning a value.

SUMMARY

In this chapter, we covered a great deal of basic information about VB .NET. We
began with variables and then moved on to constants, structures, and the CTS.
We created several examples of declaring variables and constants. After these topics,
we moved our attention to arrays and showed you how to use them. Lastly, we
looked at several varieties of loops and then briefly touched on functions and procedures. In Chapter 6, Object-Oriented Programming with VB .NET, we look at
what is arguably the most important new feature of VB .NET: its object-oriented
techniques.

6 Object-Oriented Programming with VB .NET

Since version 4, Visual Basic has offered some object-oriented (OO) programming abilities, although the implementation was less than perfect. With this release, VB .NET becomes the first truly OO version of VB. There is now full inheritance, along with other features such as abstraction, encapsulation, and polymorphism. VB .NET now even allows you to use classes written in other programming languages, such as C# or C++.

It is important to understand how object orientation will change your programs and the process of creating applications in VB .NET. It is a big shift in the way you'll write your applications and, because this is such a large topic, there are volumes of materials already written on object-oriented concepts and how they can be utilized effectively. Obviously, we can't re-create all of those works in a single chapter, so we'll focus on the ideas that most often come into play with VB .NET and how they can be used.

In this chapter, we go over each of these ideas, which may be new to VB programmers.

ABSTRACTION

VB has supported abstraction since version 4, and it is actually a simple concept. It is a view of an entity that includes only those aspects that are relevant for a particular situation. In other words, it is the ability of a language to create "black box" code that takes a concept and creates an abstract representation of that concept within a program.

For instance, suppose we want to create the code that provides services for keeping an employee information database. We'll need to store the following list of items:

- Name
- ID
- IncreaseSalary
- DecreaseSalary
- Salary

One of the most important ideas to keep in mind is that we included several items in this list for basic information, such as Name and ID. However, we also have an action entity that will be used to increase or decrease salary. These actions are referred to as methods in VB .NET. In pseudocode, the object takes the following form:

```
Employee Object
    Name
    ID
    IncreaseSalary()
    DecreaseSalary()
    Salary
End Employee Object
```

ENCAPSULATION

Encapsulation is the process of taking the abstract representation that we create and encapsulating the methods and properties, exposing only those that are totally necessary from a programmer's standpoint. The properties and methods of the abstraction are known as a member of the abstraction. The entire set of the members is known as the interface.

Simply put, this allows a developer to control the methods and properties that are available outside the object.

INHERITANCE

VB .NET is the first version of VB that supports inheritance, which is the idea that a class can gain the preexisting interface and behaviors of an existing class. In other words, a class can inherit these behaviors from the existing class through a process known as subclassing.

When an object is inherited, all of its properties and methods are automatically included in the new object. For example, suppose we look back at our Employee pseudocode object. Now, let's create a new object that tracks managers instead of employees:

```
Manager Object
    Inherits Employee Object
    ManagerPosition
End Manager Object
```

This new `Manager` object will now have the same properties as the `Employee` object (an ID, Name, etc.), but also includes a property to detail the position the manager holds. This is a very powerful way to share the code that you have already written instead of re-creating the code.

POLYMORPHISM

Polymorphism was introduced with VB4. Polymorphism simply means having or passing through many different forms. That is, it is the ability to write one routine that can operate on objects from more than one class, while treating different objects from different classes in exactly the same way. An easy example would be to create several classes that inherit the `Name` property from the `Employee` class we have been looking at. Basically, this means that the `Name` property would be available in many different forms.

Polymorphism allows an inherited method to be overridden. Again, looking at the `Employee` object, we could create a new object that inherits the `Employee` object, but with a new method with the same name as the `IncreaseSalary` method we used earlier. Although the new object would then have the methods and properties of the original, because the new method exists, it executes it.

CLASSES

A class is used to define an object. It is a sort of template that you can use to define the properties and methods of your objects and is the structure that can contain events, constants, and variables. To create a class, you begin with the `Class` keyword and assign a name to create instances of the class. Any time we need to create a class in VB .NET, we simply put all the code for the class within the `Class...End Class` block. This is similar in many ways to the loops we looked at in the Chapter 5, Introduction to the VB .NET Language.

Statements within the class comprise its methods, properties, and events. Members declared as Private are available only within the class. Public members, on the other hand, are available outside the class and are the default declaration if neither is specified.

Methods in VB .NET are created using the Sub or Function keywords. A method created with Sub does not return a value, whereas a Function must return a value as a result. We've already looked at the concepts of a Function and Sub, so we won't spend any additional time on them. You can declare methods as follows:

Private: Only visible within the class

Friend: Visible by code within the project

Public: Visible by code outside the class

Protected: Available to subclasses

Protected Friend: Visible to the project and by code in subclasses

Properties store information in an object and can be specified by either public variables or as property methods. When properties are declared using a public variable, they are also referred to as fields. This method does not provide the ability to control (validate) reading and writing a property. Property methods allow you to control read and write operations of properties in a class using the ReadOnly keyword.

For example, let's look again at the Employee object we have talked about in pseudocode throughout this chapter. We'll create a simple class:

```
Class Employee
Public ID As Integer = 1

        Public Property Salary() As Integer ' This is a Property.
        Get
        Return Salary
        End Get

        Set(ByVal Value As Integer)
        NumWheelsValue = Value
        End Set
        End Property
End Class
```

This doesn't take into account everything we would need to include in a real Employee object, such as name and so on, but the concepts for creating them are exactly the same as the previous code. We begin by creating a Public Property called

Salary and then use Get/Set to retrieve or set the property value in our code. For example, after the class is available in the project, you can create a new Employee object as follows:

```
Dim clsEmployee As New Employee()
```

Then, you could assign a salary to an employee as follows:

```
clsEmployee.Salary = 500
```

We could also add methods to the example class:

```
Class Employee
Public ID As Integer = 1

        Public Property Salary() As Integer ' This is a Property.
        Get
        Return Salary
        End Get

        Set(ByVal Value As Integer)
        NumWheelsValue = Value
        End Set
        End Property

Public Sub IncreaseSalary()
        'We need code for increasing salary
End Sub

End Class
```

Again, these are very simple examples. Don't worry if all of this doesn't click at this time. We review these types of ideas as we use them in the book in real examples.

CONSTRUCTORS

Constructors are methods of a class that are executed when a class is instantiated. They are very often used to initialize the class. To create a constructor, you simply add a public procedure called New() to your class. You can also use a parameterized constructor. This allows you to create a class that can have parameters passed to it when it is called.

Back to our employee class concept, we could create a constructor as follows:

```
Public Sub New()
    Salary = 500
End Sub
```

Using this example, when the class is initialized, we set the salary to a value of 500. This doesn't mean that it cannot be changed, but instead of assigning every employee a salary, they would begin with a salary of 500 that could be added to, subtracted from, or just changed completely.

We could also allow the user of the object to the constructor to pass in the initial value instead of assigning it to 500. We could use a parameterized constructor to do this. We add a parameter for the InitialSalary as follows:

```
Public Sub New(ByVal InitialSalary As Integer)
    Salary = InitialSalary
End Sub
```

The user of the class could then create an object as follows:

```
Dim clsEmployee as New Employee (500)
```

This works well, but suppose that we want to offer a third option for the user of the class. Instead of assigning the salary by default to a value of 500, or using a constructor that required the user to assign it, let's use the Optional keyword to create an optional constructor:

```
Public Sub New(Optional ByVal InitialSalary As Integer = 500)
    Salary = InitialSalary
End Sub
```

Now, if the user assigns a value, it overrides the default 500 value. If a value is not passed, the value is the default of 500.

Although these have been very simple examples, you can see how constructors can be utilized in your applications.

OVERLOADING

Overloading is one of the more appealing aspects of polymorphism. It allows you to create multiple methods or properties with the same name but with different parameter lists. You can change the parameters to completely different types. For instance, look at the following example Sub:

```
Public Sub MyMethod(X As Integer, Y As Integer)
```

To overload this method, you can come up with an entirely different list of parameters. For example, you can use Integer and Double:

```
Public Overloads Function MyFunction(X As Integer, Y As Double)
```

or Double and Integer:

```
Public Overloads Function MyFunction (X As Double, Y As Integer)
```

To put this into an example, let's suppose we want to create a class that multiplies numbers together. We could use the Overloads keyword to create the methods and then perform the appropriate action as necessary. Although it really doesn't make any difference in our simple example, we're going to create options for passing two integers or two real numbers. We could have made a single function that would have taken either of these, but for example purposes, we'll ignore this obvious fact:

```
Public Overloads Function Multiply(ByVal x As Integer, ByVal y As
Integer)
     Return x * y
End Function
Public Overloads Function Multiply(ByVal x As Double, ByVal y As
Double)
     Return x * y
End Function
```

Then use these:

```
clsName.Multiply(1,5)
clsName.Multiply(1.5, 5.5)
```

Now, when it is utilized, if two integers are passed (such as in the first example previously), the first function is executed. Interestingly, if one of the two parameters is a real number, the second multiply function is executed. This is because VB implicitly converts the Integer parameter to a Double data type in this instance. Similarly, if both values that are passed are real values, the second function is executed.

OVERRIDING

When you inherit a class, it allows you to use all of the methods and properties available in the class. Although this is obviously very useful, there might be times when you want to alter the inherited properties or methods. Instead of creating a new method or property with a new name, you can simply override the existing member, which is another feature of polymorphism.

To do this, you use the Overridable keyword in the original base class and then the Override keyword in the class derived from the original. For example, we'll use our original employee class:

```
Class Employee
    Public Overridable Function IncreaseSalary(byVal amt as Integer) As
Integer
        Return amt+Salary
    End Function
End Class
```

Now, let's take a look at a new class derived from this one:

```
Class Employee2
Inherits Employee
    Public Overrides Function IncreaseSalary(ByVal amt As Integer) As
Integer
        Return amt*2
    End Function
End Class
```

Although you will notice that the IncreaseSalary function that we created doesn't have any real program values, it is easy to use as an example. In the original class, we take the amount passed to the function and then add it to salary. In the inherited class, we take the same amount and double it. You will notice the Overridable and Overrides keywords in the previous code were the only additional requirements.

There are a few additional keywords that you can use when dealing with overriding members of a class. In addition to the two we already looked at (Overrides and Overridable), we can use the NotOverridable keyword to specify a method or property that cannot be overridden. You do not actually have to use NotOverridable because it is the default; so, unless you specifically use Overridable, it can't be overridden. You can choose to use NotOverridable or simply leave it blank—both are acceptable.

Another keyword is `MustOverride`. As its name indicates, if you use `MustOverride`, classes that are derived from a base class are required to override the property or method. For example, suppose we were to create a class that did mathematical calculations, but at the time of its creation, we didn't know what type of math was going to be used. We could simply create a `MustOverride` method in our class and then the classes that were derived from the original would be forced to override the method and use the appropriate math.

SHARED MEMBERS

We'll look at one final area related to classes in VB .NET: Shared Members. If you create multiple instances of a class and then change the value of a property in one of them, it does not alter the value of the other. If you have a need to simultaneously alter members of multiple classes, you can create a shared member that allows you to create members that are shared among all instances of a class.

NAMESPACES AND ORGANIZING CLASSES

As we have been working our way through the basics of VB .NET, you may have been wondering how the classes are organized. After all, if you have methods from many different classes, it's likely that you could have some names that are identical. You wouldn't want to browse through hundreds of method names just to make sure that you didn't use the same names in your classes, not to mention the classes that you will ultimately use that others have created.

For example, let's look at a very common method name such as "Open." You might have a method named "Open" for many different classes that all have unique code written for it. With this in mind, we need a way to organize the many classes and methods that are part of .NET so that they do not conflict with one another. This is where the .NET namespace comes into play.

Understanding the concept of the namespace is very simple as you have been using similar ideas in everyday life. For example, almost every town has a single or multiple McDonald's restaurant. You distinguish between the restaurants by their location. These addresses are similar to the way the namespace works in VB .NET because classes and methods in your application can be separated by locating them accordingly in a namespace.

For example, suppose we have a custom class with an `Open` method. The class is `myclasses.clayton.fill`. The built-in class `System.IO` has an `Open` method as well, related to handling the file input and output (I/O) features in VB .NET. So, if we were writing an application and wanted to include the `System.IO` features, we could do so as follows:

```
Imports System.IO
```

We could then use the class as follows:

```
Dim oFile As FileStream = New FileStream ("C:\text.txt", FileMode.Open,
FileAccess.Read)

Dim oStream As StreamReader = New StreamReader(oFile)

MsgBox(oStream.ReadLine)
oFile.Close()
oStream.Close()
```

You'll notice that in the previous code, we begin with the `Imports` keyword. We could have typed the entire `System.IO.FileStream` out everytime we wanted to use it, but instead, we can use the `Imports` keyword to allow us to shorten it to just `FileStream`. Similarly, we use `StreamReader` instead of `System.IO.StreamReader`. This makes your code much easier to read and follow while saving you time while typing. We're not really interested at this time in what the previous code does, but are just using it as an example of how classes are organized and how the `Imports` keyword can be used in your future applications.

SUMMARY

In this chapter, we covered the basics of object-oriented programming in VB .NET. We touched on abstraction, encapsulation, inheritance, polymorphism, classes, constructors, overloading, overriding, shared members and namespaces, and organizing classes. In Chapter 7, we look at strings, GDI +, and error handling.

7 Strings, GDI+, and Error Handling in VB .NET

As you have seen in the previous chapters leading up to this point, there are many changes that have taken place in this version of VB. There are completely new approaches to development, and although some of the old functions are still in place, it is recommended that you focus on learning the new items in VB .NET rather than on functions that may or may not exist down the road.

In this chapter, we continue to look at some of these new items, and specifically deal with the changes to strings, graphics, and error handling in VB .NET.

.NET STRINGS

If you have been using previous versions of VB, you will have undoubtedly utilized the many powerful string functions it has always offered. Like others areas of VB .NET, the string functions are still there, but they are implemented in a different manner. Now, if you remember that everything in VB .NET is considered an object (although this has been repeated many times, it is important to remember this), you will realize that as you Dim a string variable, you are actually creating an instance of the String class.

If you refer back to Chapter 5, Introduction to the VB .NET Language, in which we looked at some of the basics of the .NET language, you will see that we did a little work with strings and how they are used. We expand on that information in this chapter, beginning with the list of the common methods (see Table 7.1) that are built-in to the String class.

A quick glance at Table 7.1 is all that it takes to see some of the ways you can use the various methods that are available in .NET. Before we look at a specific example, you should be aware that the index of the first character in a VB .NET string is a zero rather than a one. In previous versions of VB, strings were one-based instead of zero-based. Although simple, it's something that you have to keep in mind as you are upgrading projects or using VB .NET to build new ones, because the old habits occasionally creep into the picture.

TABLE 7.1   Common methods built-in to the String class

Class Method	Description
Compare	Compares two strings
Concat	Concatenates strings
Copy	Creates a new instance with the same value
Equals	Determines if two strings are equal
Format	Formats a string
Equality Operator	Allows strings to be compared
Chars	Returns the character at a specified position in string
Length	Returns the number of characters in string
EndsWith	Checks string to see if it ends with a specified string
IndexOf	Returns the index of the first character of the first occurrence of a substring within this string
IndexOfAny	Returns the index of the first occurrence of any character in a specified array of characters
Insert	Inserts a string in this string
LastIndexOf	Returns the index of the first character of the last occurrence of a substring within this string
LastIndexOfAny	Returns the index of the last occurrence of any character in a specified array of characters
PadLeft	Pads the string on the left
PadRight	Pads the string on the right
Remove	Deletes characters at a specified point
Replace	Replaces a substring with another substring
Split	Splits a string up into a string array
StartsWith	Checks if a string starts with the specified string
SubString	Returns a substring within the string
ToLower	Converts string to all lowercase letters
ToUpper	Converts string to all uppercase letters
Trim	Removes all specified characters from string
TrimEnd	Removes all specified characters from end of string
TrimStart	Removes all specified characters from beginning of string

Here is a simple example using the string methods:

```
Dim strTest as String = "ABCDEFG"
Dim strlength as Integer

strlength =  strTest.Length()
```

This example simply takes the string variable `strTest` and assigns the length of it to `strlength`.

We can test other features simply as well:

```
strTest = strTest + "HIJKLMNOP"
```

Now, the string that is stored in `strTest` is actually "ABCDEFGHIKLMNOP".

We use strings in many of the sample applications we develop later, so we're going to focus our attention on the new and exciting graphics capabilities offered in VB .NET.

GRAPHICS WITH GDI+

In VB .NET, we have a completely new way of drawing graphics. It is based on drawing to a Graphical Device Interface (GDI) surface rather than using the old line, circle, and print methods. If you have used C++ in the past, you know the frustration that was involved in GDI development with MFC. And if you are a VB developer, you know how limited you have been with the simple methods that were available in VB6 and earlier.

In VB .NET, we now have access to GDI+. It's a superior implementation and is easier to use than traditional GDI. GDI+ provides a variety of functions. For example, if you want to set the background or foreground color of a control, you can set the `ForeGroundColor` property of the control.

GDI+ is easier to use, but there are also many new features that have been added, including:

- Antialiasing support
- Gradient brushes
- Splines
- Transformation and matrices
- Alpha blending

There are several namespaces that you need to become comfortable with to master GDI+ development. The GDI+ classes are grouped under six namespaces, which reside in the `System.Drawing.dll` assembly. These namespaces include `System.Drawing`, `System.Drawing.Design`, `System.Drawing.Printing`, `System.Drawing.Imaging`, `System.Drawing.Drawing2D`, and `System.Drawing.Text`.

Let's take a quick look at a few of these namespaces.

System.Drawing **Namespace**

The `System.Drawing` namespace provides the basic GDI+ functionality. It contains the definition of basic classes, such as `Brush`, `Pen`, `Graphics`, `Bitmap`, `Font`, and so on. The `Graphics` class plays a major role in GDI+ and contains methods for drawing to the display device. Table 7.2 details some of the `System.Drawing` namespace classes and their definitions. Table 7.3 describes the namespace structures and their definitions.

TABLE 7.2   System.Drawing namespace classes

Class	Description
Bitmap	Bitmap class
Image	Image class
Brush (Brushes)	Defines objects to fill GDI objects, such as rectangles, ellipses, polygons, and paths
Font (FontFamily)	Defines particular format for text, including font face, size, and style attributes
Graphics	Encapsulates a GDI+ drawing surface
Pen	Defines an object to draw lines and curves
SolidBrush (Texture Brush)	Fills graphics shapes using brushes

TABLE 7.3   System.Drawing namespace structures

Structure	Description
Color	RGB color
Point (PointF)	Represents 2D x and y coordinates. Point takes x, y values as a number. If you need to use floating-point numbers, you can use PointF.

TABLE 7.3   `System.Drawing` namespace structures (*continued*)

Structure	Description
Rectangle (RectangleF)	Represents a rectangle with integer values for top, left, bottom, and right. RectangleF allows you to use floating-point values.
Size	Defines size of a rectangular region given in width and height. SizeF takes floating numbers.

`System.Drawing.Drawing2D` Namespace

This namespace contains 2D and vector graphics functionality. It contains classes for gradient brushes, matrix and transformation, and graphics paths. Some of the common classes are shown in Table 7.4. Table 7.5 displays the common enumerations.

TABLE 7.4   `System.Drawing.Drawing2D` namespace classes

Class	Description
Blend (ColorBlend)	Defines the blend for gradient brushes
GraphicsPath	Represents a set of connected lines and curves
HatchBrush	Provides a brush with hatch style, a foreground color, and a background color
LinearGradientBrush	Provides functionality for linear gradient
Matrix	Represents geometric transformation in a 3x3 matrix

TABLE 7.5   `System.Drawing.Drawing2D` namespace enumerations

Enumeration	Description
DashStyle	Represents the style of dashed lines drawn with the Pen
HatchStyle	Represents different patterns available for the HatchBrush
QualityMode	Specifies the quality of GDI+ objects
SmoothingMode	Quality (smoothing) of GDI+ objects

GRAPHICS CLASS

The Graphics class is a key component of GDI+ development. Before you draw an object, such as a rectangle or line, you need a surface to draw it to. You must use the Graphics class to create these surfaces before you can draw.

There are a few ways in which you can get a graphics object to use in your application. First, you can get a graphics object in the form paint event. Another option is to override the OnPaint() method of a form. Either way, you use System.Windows.Forms.PaintEventsArgs.

Here is a simple example of overriding the OnPaint() method:

```
protected overrides sub OnPaint(ByVal e As
System.Windows.Forms.PaintEventArgs)

    Dim g As Graphics = e.Graphics

End Sub
```

This creates the graphics object for us and we can now use any of the methods that are included with the Graphics class. For example, we can use any of the following:

DrawArc: Draws an arc

DrawBezier (DrawBeziers): Draws Bezier curves

DrawCurve: Draws a curve

DrawEllipse: Draws an ellipse or circle

DrawImage: Draws an image

DrawLine: Draws a line

DrawPath: Draws a path

DrawPie: Draws outline of a pie section

DrawPolygon: Draws outline of a polygon

DrawRectangle: Draws outline of a rectangle

DrawString: Draws a string

FillEllipse: Fills the interior of an ellipse

FillPath: Fills the interior of a path

FillPie: Fills the interior of a pie section

FillPolygon: Fills the interior of a polygon defined by an array of points

FillRectangle: Fills the interior of a rectangle

FillRectangles: Fills the interiors of a series of rectangles

FillRegion: Fills the interior of a region

Now that we have a basic understanding of some of the things we can do, let's try a simple example to draw a line:

```
protected overrides sub OnPaint(ByVal e As
System.Windows.Forms.PaintEventArgs)

    Dim g As Graphics = e.Graphics
    Dim pn As Pen = New Pen(Color.Green, 5)
    g.DrawLine(pn, 1, 50, 100, 500)

End Sub
```

OBJECTS

You have access to several types of objects when drawing items such as an ellipse and a line. There are four common GDI+ objects, detailed in the following list, which you can use to fill GDI+ items.

Brush: Used to fill enclosed surfaces with patterns, colors, or bitmaps

Pen: Used to draw lines and polygons, including rectangles, arcs, and so on

Color: Used to render a particular object (in GDI+, color can be alpha blended)

Font: Used to render text

We're going to look at a few examples of each of these.

The Brush Class

The Brush class is an abstract base class and as such, we always use its derived classes such as SolidBrush, TextureBrush, RectangleBrush, and LinearGradientBrush to instantiate a brush object.

Here is an example:

```
Dim brsh as SolidBrush = new SolidBrush(Color.Red), 40, 40, 140, 140
```

The Pen Class

A Pen, as you have already seen, draws a line of specified width and style. You can initialize a new instance of the Pen class with the specified color, a specified brush, a specified brush and width, or a specified color and width. Here is an example:

```
Dim pn as Pen = new Pen( Color.Blue )
```

The previous code initialized the Pen with the color blue. The following example initializes it and also assigns a width of 10:

```
Dim pn as Pen = new Pen( Color.Blue, 10 )
```

Color Structure

A Color structure represents an ARGB color. ARGB properties are described in the following list.

A: Alpha component value for the color

R: Red component value for the color

G: Green component value for the color

B: Blue component value for the color

You use the Color structure to change the color of an object and you can call any of its members. For example, you can change the color of a Pen as follows:

```
Dim pn as Pen = new Pen( Color.Red )
```

or

```
Dim pn as Pen = new Pen(Color.Green)
```

The Font Class

The Font class defines a particular format for text, such as font type, size, and style attributes. You can create a new instance of the Font class as follows:

```
Public Sub New(Font, FontStyle)
```

Here is an example:

```
Dim myFnt as Font = new Font ("Times New Roman",12)
```

There are many variations and possibilities of what you can do with the Font class, and we look at them more closely as we use them in the examples in this book.

ERROR HANDLING

In earlier versions of VB, we used the On Error GoTo statement to handle errors. This traditional error checking in VB has involved checking the result of a function and then acting appropriately to whatever value is returned. Although checking errors in this manner has worked, its effectiveness is often a gamble. For example, you could check the value of an incorrect value or even forget one of the values that causes an error. This leaves this type of error checking open to many more problems. There are times when you will be forced to do this, but for many errors in VB .NET, we can use exception error handling.

To perform this type of error handling at execution, we check for runtime errors that are called exceptions. When an error is encountered, the program takes the appropriate steps to continue executing. The code that is executed will only be executed when an error has occurred. We use the Try and Catch keywords in this form of error handling, which typically works as follows:

```
Try
      Code
Catch
      Error
End Try
```

A simple way to dissect this approach is to look at the keywords. For starters, you can consider that the Try keyword is used to begin the exception handler. You place code inside this section that you believe could cause some problems. The next section, Catch, is the executed code that tries to help your program overcome the errors. You can use multiple Catch blocks of code if you need different code for different errors. For example, suppose you have some code that is written to save a file of some type. The error handling can be present for several types of errors:

- The user didn't enter a filename.
- An invalid character was entered for the filename, such as "*".
- A network drive that was previously existent is no longer available.
- And so forth . . .

You can create multiple Catch blocks that handle these errors appropriately—as you probably don't want your application responding to an invalid filename in the same way it responds to an unavailable network drive.

Here is an actual example that causes a "Division by Zero" error:

```
Dim a as Integer = 10
Dim b as Integer = 0
Dim c as Integer = 0

Try
      b = a / c
Catch
    MessageBox.Show(err.ToString)
      b=0
End Try
```

In the previous code, we begin with three integer variables: a, b, and c. The variable b is going to hold the value of a divided by c. Because we initialize c with a value of 0, the attempted division is met with an error. The line is inside the Try block of code, so it then moves on to the Catch block for handling the error. We display a message box that details the error and then sets the variable b equal to 0 so that the program can continue execution.

At this time, the code sets b equal to 0 regardless of the program error. We can further enhance this example by checking for the specific Divide by Zero error as follows:

```
Dim a as Integer = 10
Dim b as Integer = 0
Dim c as Integer = 0

Try
      b = a / c
Catch As DivideByZeroException
    MessageBox.Show("You tried to divide by zero.")
      b=0
    Catch err As Exception
        MessageBoxShow(err.ToString)
End Try
```

This example checks the DivideByZeroException and if the code encounters it, it executes the appropriate code. Otherwise, it continues on and the next Catch finds any other errors that are generated.

SUMMARY

As we have done in previous chapters, we covered a great deal of information beginning with .NET strings. From strings, we ventured into graphics with GDI +, the `Graphics` class, objects, and touched upon some of the error handling capabilities of VB .NET. In Chapter 8, Math and Random Number Functions in VB .NET, we cover .NET math.

8 Math and Random Number Functions in VB .NET

In previous versions of VB, there were a multitude of functions that aided a developer in the creation of complex mathematical development. These functions also exist in VB .NET, but like everything else, they are implemented a little differently and reside in the System.Math class.

THE System.Math CLASS

The System.Math class contains a variety of methods that you can use for mathematical calculations. There are functions such as square root, pi, absolute value, rounding, and trigonometry to name a few. We look at each of them with a little detail here, and later in the book, we use them to build a fully functional calculator.

Raising to a Power

We begin our discussion of the System.Math class by looking at the Pow method. The Pow method is used to raise a number to a power. For example, you can raise a number to a power of 2 with the following code:

```
Dim X As Integer
X = Math.Pow(2, 2)
```

The previous code is assuming that we imported the System.Math class. We then assign the integer variable X equal to 2 to a power of 2. We can also rewrite the code as follows:

```
Dim X As Integer
X = 2 ^ 2
```

Now, let's use the Pow method to perform a useful calculation. For example, suppose we are interested in calculating the area of a circle (it is equal to pi * radius squared). We can do this as follows:

```
Dim X as Double
Dim dblRadius as Double

dblRadius = 50
X = Math.PI * Math.Pow(radius,2)
```

The previous code assigns a value of 50 to the radius and then calculates the area and assigns it to X. You'll notice that we used Math.PI for the value of pi. It is another built-in method available to us in VB .NET.

Square Root

The square root method (Sqrt), like the other VB .NET math functions, resides in the System.Math class. You can use it to calculate the square root of a value. The following example demonstrates its use:

```
Dim X as Double
X = Sqrt (100)
```

Absolute Value

We can use the absolute value (Abs) method to return an absolute value of a number. If you are unfamiliar with the absolute value, it is simply the value of a number, without regard to its sign. In other words, it is a positive number. Here is an example:

```
Dim X as Double

X = Abs( -10.5)
```

This returns a value of 10.5. Similarly, the following example returns the same value:

```
Dim X as Double

X = Abs(10.5)
```

However, if you want to return the sign of a number, you can use the Sign method.

If the number is negative, Sign returns -1; if it's positive, Sign returns 1; and if the number is equal to 0, Sign returns 0. Here is an example:

```
Dim X1, X2, X3 As Double
Dim X4 As Integer
X1 = 5
X2 = -5
X3 = 0
X4 = Sign(X1)
```

This returns a value of 1.

```
Dim X1, X2, X3 As Double
Dim X4 As Integer
X1 = 5
X2 = -5
X3 = 0
X4 = Sign(X2)
```

This returns a value of -1.

```
Dim X1, X2, X3 As Double
Dim X4 As Integer
X1 = 5
X2 = -5
X3 = 0
X4 = Sign(X3)
```

This returns a value of 0.

Rounding Numbers

If you need to round values to the nearest integer value, you can use the Round method. As an example, suppose you have a value of 5.12345 and need to round it:

```
Dim X as Double
X = Math.Round(5.12345)
```

This rounds it to 5. There are a few things to remember when using the Round method. If you have a number that is between two numbers (such as 5.5 or 6.5), the method may not return what you would at first believe. Most people would round these values up to the next highest value; however, if you create the following code, something interesting happens:

```
Dim X1, X2 As Double
X1 = Math.Round(5.5)
X2 = Math.Round(6.5)
```

The values of X1 and X2 both return values of 6. This is because the Round method actually returns the EVEN number closest to the two. To truncate a number in VB .NET, you can use the Floor method. It returns the largest whole number smaller than the orginal number. So, the following code returns 5:

```
Dim X as Double
X = Math.Floor(5.6)
```

Negative values can behave unexpectedly, so you have to pay special attention to them. For example, if you create a similar project with a negative value, it returns a -6:

```
Dim X as Double
X = Math.Floor(-5.6)
```

Trigonometry

The Math class also contains a number of methods for making trigonometric and logarithmic calculations. These include methods for Sin, Cos, Tan, and Atn. They work like the other methods we have been looking at throughout this chapter:

```
Dim X as Double
X = Sin(1.1)
```

This returns a value in radians. If you want to convert from radians to degrees, you multiply by 180/pi:

```
Dim X As Double
X = 180 / Math.PI * Sin(1.1)
Debug.WriteLine(X)
```

If you have a degree value, you can convert it to radians by multiplying by pi/180.

Logarithms

The System.Math class also provides functionality for logs and natural logs. You can use the Exp method of the class to return e raised to a power:

```
Dim X as Double
X = Math.Exp(2)
```

You can also use the `Log` method of the `Math` class to return the natural logarithm of a number:

```
Dim X as Double
X = Math.Log(5)
```

CREATING YOUR OWN MATH FUNCTIONS

There are obviously many great methods related to math built into the `System.Math` class. However, there are going to be times when you need to create your own. We can take a simple example for conversions to see how and why you will do so. For example, let's suppose you are trying to convert Celsius to Fahrenheit. The formulas for conversion are as follows:

Celsius to Farhenheit: `Value * 1.8 + 32`

Farhenheit to Celsuis: `(Value - 32)/1.8`

The functions can then be created as follows:

```
Function CToF(ByVal value as Single) as Single
    CToF = value * 1.8  +32
  End Function

Function FToC(ByVal value as Single) as Single
    FToC=(value-32)/1.8
End Function
```

We can use the functions in a program as follows:

```
Dim X As Single
X = CToF(25)
```

As an additional example, suppose we need to return the decimal part of a number. We can calculate this simply by subtracting the number from the decimal portion:

```
Function Decm(Value as Double) As Double
    Decm = value - fix(value)
  End Function
```

We could then use this as follows:

```
Dim X as Double
X = Decm(10.5)
```

This returns a value of 0.5.

GENERATING RANDOM NUMBERS

The System.Random class is used to draw a random number, and unlike the VB6 Rnd function, System.Random can return both decimal and whole random numbers. Additionally, unlike Rnd, System.Random automatically seeds its random number generator with a random value derived from the current date and time.

You can use the NextDouble method of System.Random to return a Double random number between 0 and 1. You can use the Next method to return an Integer random number between two integer values.

```
Dim X As Double
Dim rnd As System.Random = New System.Random()
Dim i As Integer

For i = 1 To 10
    X = Round(rnd.NextDouble() * 10)
    Debug.WriteLine(X)
Next
```

This code displays a series of random numbers between 0 and 10. You can also use the Next method as follows:

```
Dim X As Double
Dim rnd As System.Random = New System.Random()
Dim i As Integer

For i = 1 To 10
    X = rnd.Next(0, 10)
    Debug.WriteLine(X)
Next
```

SUMMARY

In this chapter, we worked on several topics that are related to the math functions available in VB .NET. This is the final chapter that consists of only text information. In most of the remaining chapters, we build applications. In doing so, we put some of the information that we have already learned to good use and also learn about many new topics. In Chapter 9, Your First Program, we create our first VB .NET application.

9 Your First Program

Now that we have walked through some of the changes in VB .NET, we're going to take some of what we have learned and put it into a simple "Hello World!" application. We're going to create a Console application that simply opens and displays the "Hello World!" text on the screen. This is by far the easiest application we create in the book, and although it is very simple, it is a good way to become acquainted with the IDE and how to build a VB .NET program.

ON THE CD

The source code for the projects are located on the CD-ROM in the PROJECTS folder. You can either type them in as you go or you can copy the projects from the CD-ROM to your hard drive for editing.

USING THE IDE

At this time, it is assumed that you have a version of Visual Basic .NET installed on your PC and that you have set the profile information as directed in Chapter 2, Introduction to Visual Basic .NET. Although it is not a requirement, the profile settings in the first chapter allow your display and the figures in this chapter to be very similar.

The first step is to open the Visual Studio .NET IDE, which can be seen in Figure 9.1.

At this time, the IDE should be at the Start page, where you can select New Project. Alternately, you can choose File | New Project. Either way, the New Project window is displayed (see Figure 9.2).

You need to assign a location for the file to be stored along with a name for your project. These items can be set however you want, but something similar to the following works well:

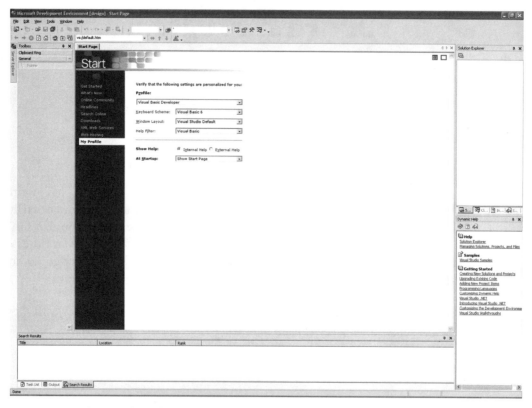

FIGURE 9.1 The Visual Studio .NET IDE.

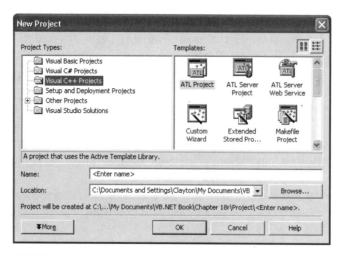

FIGURE 9.2 New Project window in the IDE.

Name: Chapter8

Location: C:\VBNET\Projects\

You will notice a variety of application templates that can be used in VB .NET. At the left of the window, notice the Project Types list. Make sure that Visual Basic is selected from the list as it is in Figure 9.3.

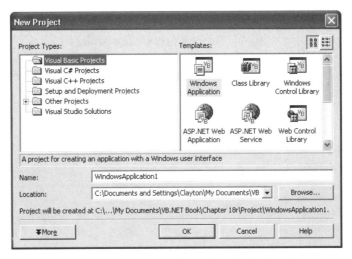

FIGURE 9.3   VB needs to be selected as the Project Type.

The following list describes the various templates:

Windows Application: A traditional standalone Windows application

Class Library: A windowless project that is a reusable class or component that can be shared with other projects

Windows Control Library: A custom control to use on Windows Forms

ASP.NET Web Application: The creation of an ASP.NET Web application

ASP.NET Web Service: The XML Web services authored with ASP.NET that can be published and called by an external application

Web Control Library: A custom control that can be used on Web Forms pages (similar to ActiveX control creation in VB6)

Console Application: A command-line application—we are using this in our program

Windows Service: The applications that do not have a user interface; formerly called "NT services," these applications are used to do things such as monitor files or check performance of the machine

Empty Project: An empty project with the necessary file structure to store application information, but all references, files, or components must be added manually

Empty Web Project: An empty environment for server-based applications

New Project in Existing Folder: A blank project within an existing application folder for using files from a preexisting project

For our application, we are going to choose Console Application. After setting the name and file location and clicking Console Application, click the OK button. Your screen should now look similar to Figure 9.4.

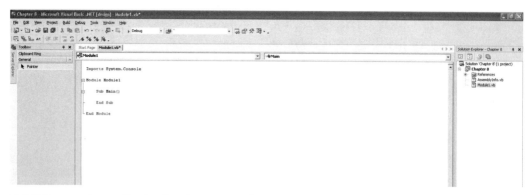

FIGURE 9.4   The application has been created.

We are now in Module1.vb. Notice the Toolbox on the left side of the IDE and Solution Explorer on the right. Because this is such a simple application, we don't have a need for either of these right now, but you can note where they are located as we'll use them in most of the later examples.

WRITING CODE

We are now in a good position to write some code for our application. This consists of a single line of output to the Console window that displays the words "Hello

World!" First, we'll perform this in a single line, then we'll accomplish it using a variable, and lastly, we'll achieve it by using the Imports keyword. All of them are equally effective for this application, but it is useful to compare how you can perform various actions in VB .NET.

Using a Single Line

Our first attempt at this application is to write a single line of code to output to the Console window. You will see that the current VB created code is similar to the following:

```
Module Module1

    Sub Main()

    End Sub

End Module
```

Sub Main() is the entry point for our application. We must write our output line between Sub Main () and End Sub. We are going to use the Systems.Console class to display information to the window. First, click in the Code Editor so that your cursor is positioned between the Sub Main() and End Sub code. Next, add the following text:

```
System.
```

Notice that as you type, Visual Studio helps you with the names of classes and functions (see Figure 9.5), because the .NET Framework publishes the type information. You can continue adding characters to the line as follows:

```
System.Cons
```

At this time, you will see System.Console selected. You can continue typing the rest of the word "Console," or simply press the Tab key on your keyboard to move to the next item.

Continue adding the following code:

```
System.Console.WriteLine
```

Again, you can type out the entire line, or let Visual Studio fill it in for you and then press the Tab key. Either way, when you have finished typing the line, press the spacebar, which brings up the parameter list for the class (see Figure 9.6).

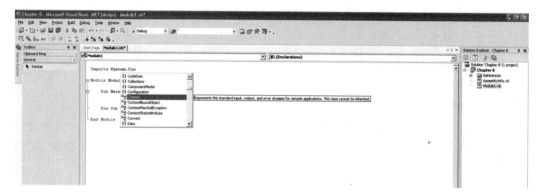

FIGURE 9.5 Visual Studio helps you with the names of available classes and functions.

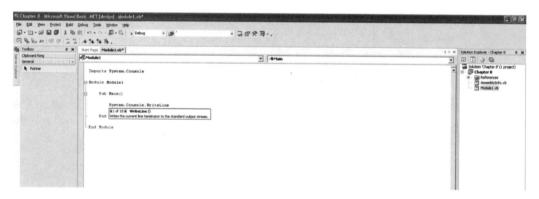

FIGURE 9.6 The parameter list for `System.Console.WriteLine`.

You can scroll through the list using the arrow keys; you can also see a few of the options in Figures 9.7, 9.8, and 9.9.

The list continues on, and in this case, there are 18 of them. You don't have to scroll through each of the items unless you choose to do so. It is a great feature if you don't know or remember the various parameters for a class or function.

We simply need to complete the line as follows:

```
System.Console.WriteLine("Hello World!")
```

Our complete code listing should look like this:

```
Module Module1
```

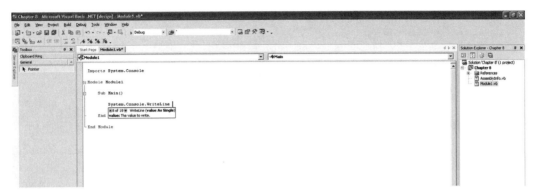

FIGURE 9.7   The first parameter list.

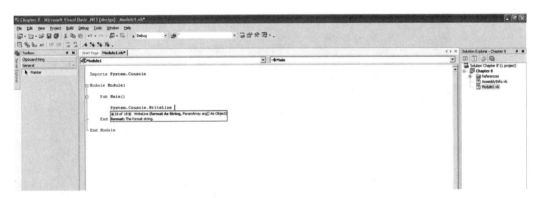

FIGURE 9.8   A second type of parameter.

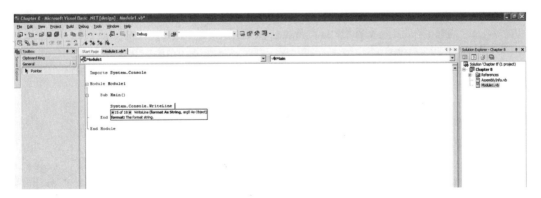

FIGURE 9.9   A third parameter.

```
Sub Main()
    System.Console.WriteLine("Hello World!")
End Sub

End Module
```

Executing the Program

We are now ready to execute the program in the IDE to see if it works. You can choose Debug | Start Without Debugging or press Ctrl+F5 as a keyboard shortcut. This starts the program and a window similar to Figure 9.10 is displayed on your screen.

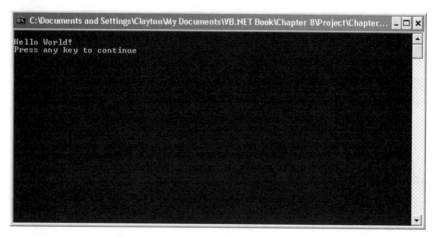

FIGURE 9.10   The "Hello World!" application.

That's all there is to this first application. If you press any key on your keyboard, you return to the IDE. Now, we're going to implement this in a slightly different way using a simple variable so that you can experience a few different features in VB .NET.

Using a Variable

We've discussed variables and declaring them in previous chapters, so we don't have to spend a great deal of time on the concepts in this chapter. We're going to create a variable of type String and then assign the variable a value. Then, we'll use the System.Console class to display this information as we did in the first example.

Begin by removing the System.Console.WriteLine("Hello World!") line from the Code Editor.

```
Module Module1

    Sub Main()

    End Sub

End Module
```

We're now back to our original code listing. If you prefer, you can start a new project instead of removing the code, but because it is only a single line, it's probably unnecessary.

We'll begin by adding the following variable declaration to the code:

```
Dim str as String
```

This code should be placed after Sub Main() so that the listing looks like this:

```
Module Module1

    Sub Main()
        Dim str As String

    End Sub

End Module
```

The next step is to assign a value to the str variable:

```
str = "Hello World!"
```

The code should now look like this:

```
Module Module1

    Sub Main()
        Dim str As String
        str = "Hello World!"

    End Sub

End Module
```

Lastly, we use the same `WriteLine` method of the `System.Console` class to output the variable information:

```
System.Console.WriteLine(str)
```

The final listing is as follows:

```
Module Module1

    Sub Main()
        Dim str As String
        str = "Hello World!"
        System.Console.WriteLine(str)
    End Sub

End Module
```

Testing the Program

Again, you can test the functionality of the program by choosing Debug | Start Without Debugging (Ctrl+F5 is the shortcut to this function).

Your output should look like Figure 9.10.

Using the `Imports` Keyword

Both of the previous options have worked; we are now going to look at a third option. This time, we use the `Imports` keyword to import the `System.Console` class into our application. This imports all of the members of the `System.Console` namespace into our application. Although we only have one line that uses it in this example, you can see how this will very effectively save us time when we are using much longer examples.

You can add the following `Imports` line above `Module Module1` in the code:

```
Imports System.Console
```

The code then looks like the following:

```
Imports System.Console

Module Module1

    Sub Main()
        Dim str As String
        str = "Hello World!"
```

```
        System.Console.WriteLine(str)
    End Sub

End Module
```

We can then shorten the `System.Console.WriteLine(str)` line of code:

```
System.Console.WriteLine(str)
```

to:

```
WriteLine(str)
```

The final code should now read:

```
Imports System.Console

Module Module1

    Sub Main()
        Dim str As String
        str = "Hello World!"
        WriteLine(str)
    End Sub

End Module
```

If you run the application again, your output will look like Figure 9.10.

FINAL CODE LISTING

The following code listings are the final listings for the three examples in this chapter:

Listing 9.1   Single Line Example.

```
Module Module1

    Sub Main()
        System.Console.WriteLine("Hello World!")
    End Sub

End Module
```

Listing 9.2 Variable Example.

```
Module Module1

    Sub Main()
        Dim str As String
        str = "Hello World!"
        System.Console.WriteLine(str)
    End Sub

End Module
```

Listing 9.3 Imports Keyword Example.

```
Imports System.Console

Module Module1

    Sub Main()
        Dim str As String
        str = "Hello World!"
        WriteLine(str)
    End Sub

End Module
```

SUMMARY

In this chapter, we built our first VB .NET application. Although it is a simple example, it is a good basis for our activity in Chapter 10, Console Application Input/Output, which is to design a Console application that takes user input and provides output. It will take the input of two numbers, calculate the third number, and then output it to the Console window.

10 Console Application Input/Output

In the previous chapter, we spent some time putting together a very simple Console application that displayed "Hello World!" when executed. We take what we learned in the example, and apply it to a new application that adds numbers together after being input by the user. Then, we add the ability to calculate the sine of a number to show the basics of the trigonometry functions being used in VB .NET.

ON THE CD

The source code for the projects are located on the CD-ROM in the PROJECTS folder. You can either type them in as you go or you can copy the projects from the CD-ROM to your hard drive for editing.

GETTING STARTED

To start, open Visual Basic and then choose Console Application from the Templates list (see Figure 10.1).

Next, we're going to add some code to the project.

Writing Some Code

The first line of code involves adding the `Imports System.Console` line to the application. This line must be present before `Module Module1`; your code should look like the following listing after you enter it:

```
Imports System.Console
Module Module1

    Sub Main()

    End Sub

End Module
```

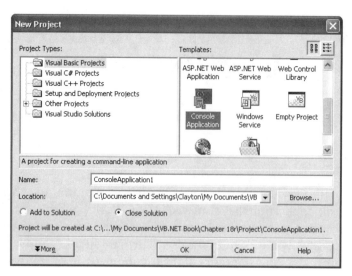

FIGURE 10.1 Choose Console Application from the Templates list.

Next, we need to declare a couple of variables for the application:

```
Dim X,Y As Double
```

These variables are used to store the values that are being input by the user and should be placed after the `Sub Main()` line in the Code Editor. The code should look like the following list:

```
Imports System.Console
Module Module1

    Sub Main()
        Dim X, Y As Double
    End Sub

End Module
```

Now, we're at a point to begin handling the input, but first, we need to let the user know that we are requesting information from them. We'll begin by using a `Write()` method to output text information, such as "Please Enter First Number:". Instead of `WriteLine()`, which actually creates an entire line with a carriage return, we'll use `Write()` so that we can leave our cursor immediately beyond the end of the line for the user input.

Here is the code with the `Write()` method added:

```
Imports System.Console
Module Module1

    Sub Main()
        Dim X, Y As Double
        Write("Please Enter First Number: ")
    End Sub

End Module
```

Next, we're ready to handle the input from the user and store it in variable X. We can use the `ReadLine()` method to handle this input:

```
Imports System.Console
Module Module1

    Sub Main()
        Dim X, Y As Double
        Write("Please Enter First Number: ")
        X = ReadLine()
    End Sub

End Module
```

We can repeat this same approach for output and input for variable Y:

```
Imports System.Console
Module Module1

    Sub Main()
        Dim X, Y As Double
        Write("Please Enter First Number: ")
        X = ReadLine()
        Write("Please Enter Second Number: ")
        Y = ReadLine()
    End Sub

End Module
```

The final step for this section of the code is to add the numbers together and then display the output using the `Write()` and `WriteLine()` methods. We're using both methods because we'll use `Write()` to display text such as "Your final answer is:", and then we'll use `WriteLine()` to actually display the answer:

```
Imports System.Console
Module Module1

    Sub Main()
        Dim X, Y As Double
        Write("Please Enter First Number: ")
        X = ReadLine()
        Write("Please Enter Second Number: ")
        Y = ReadLine()
        Write("Your final answer is: ")
        WriteLine(X + Y)
    End Sub

End Module
```

If you run the application using Debug | Start Without Debugging, you'll see output similar to Figure 10.2.

FIGURE 10.2 The code being executed.

We can check the input and output by entering values as prompted. Provide a value of 10 and then press the Enter key on the keyboard. Your window should now look like Figure 10.3.

Enter a value of 20 and then press the Enter key. This moves the program through the input, and then it displays the final answer of 30, as seen in Figure 10.4.

Because we used a Double as the type of variable, you can also use decimal point values in the program. Try the following combinations:

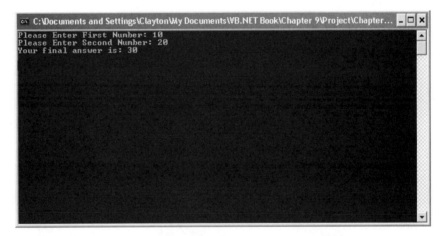

FIGURE 10.3   The program is asking for additional input.

FIGURE 10.4   The answer is displayed.

X: -1.5

Y: 1.5

X: 1.2345

Y: -55.3433

X: 3.75

Y: 4.25

The values should all work equally and aren't dependent on positive or negative values.

TRIGONOMETRY CALCULATIONS

The final step in this program is to add one additional calculation. This time, we'll leave the addition operation as is, but we're going to add input and output capabilities to calculate the sine of a number.

This requires the same ReadLine(), Write(), and WriteLine() methods that we used before:

```
Imports System.Console
Module Module1

    Sub Main()
        Dim X, Y As Double
        Write("Please Enter First Number: ")
        X = ReadLine()
        Write("Please Enter Second Number: ")
        Y = ReadLine()
        Write("Your final answer is: ")
        WriteLine(X + Y)

        WriteLine()
        Write("Calculate Sine of what number: ")
        X = ReadLine()
    End Sub

End Module
```

Notice that we are utilizing the same variable of x for reading the number. The next step, then, is to take the value stored in x and assign x equal to the sine of the value using the Sin method:

```
Imports System.Console
Module Module1

    Sub Main()
        Dim X, Y As Double
        Write("Please Enter First Number: ")
        X = ReadLine()
```

```
        Write("Please Enter Second Number: ")
        Y = ReadLine()
        Write("Your final answer is: ")
        WriteLine(X + Y)

        WriteLine()
        Write("Calculate Sine of what number: ")
        X = ReadLine()
        X = System.Math.Sin(X)
    End Sub

End Module
```

Lastly, we'll use the Write() and WriteLine() methods to display the output:

```
Imports System.Console
Module Module1

    Sub Main()
        Dim X, Y As Double
        Write("Please Enter First Number: ")
        X = ReadLine()
        Write("Please Enter Second Number: ")
        Y = ReadLine()
        Write("Your final answer is: ")
        WriteLine(X + Y)

        WriteLine()
        Write("Calculate Sine of what number: ")
        X = ReadLine()
        X = System.Math.Sin(X)
        Write("The Sine is: ")
        WriteLine(X)
    End Sub

End Module
```

If you run the program at this time, it prompts you for the first two values and then it displays the result after they have been added together. Then, after the calculation, it prompts you to enter another value. The program then outputs the sine of the value (see Figure 10.5).

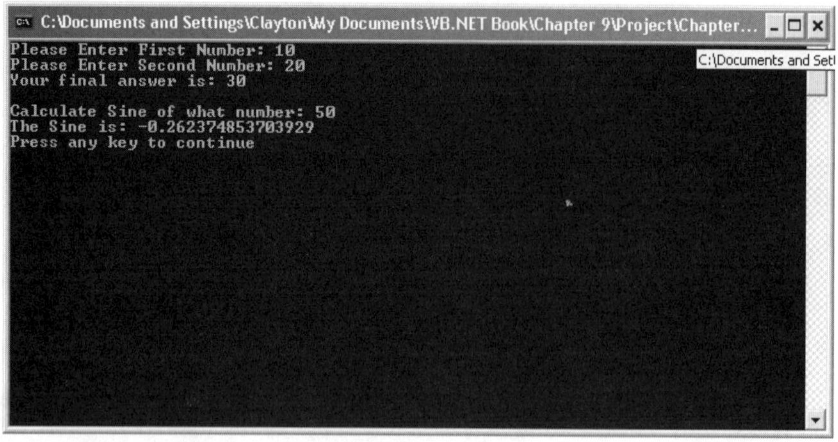

FIGURE 10.5   The final program is being executed.

FINAL CODE LISTING

This is the final code listing for the chapter:

```
Imports System.Console
Module Module1

    Sub Main()
        Dim X, Y As Double
        Write("Please Enter First Number: ")
        X = ReadLine()
        Write("Please Enter Second Number: ")
        Y = ReadLine()
        Write("Your final answer is: ")
        WriteLine(X + Y)

        WriteLine()
        Write("Calculate Sine of what number: ")
        X = ReadLine()
        X = System.Math.Sin(X)
        Write("The Sine is: ")
        WriteLine(X)
    End Sub

End Module
```

SUMMARY

In this chapter, we created another Console application; however, unlike the previous chapter, we responded to user input, and then made a few simple calculations based on the values that were entered. We used the ReadLine() method of the System.Console class to capture the input values, and the Sin() method of the System.Math class to calculate the sine of the number that was being entered. In Chapter 11, Your First Windows Forms Application, we build our first Windows Forms application.

11

Your First Windows Forms Application

I n the previous chapters, we built Console applications in VB .NET. In this chapter, we look at building a Windows Forms application, a much more common type of program for VB developers (see Figure 11.1) especially for applications related to a Tablet PC.

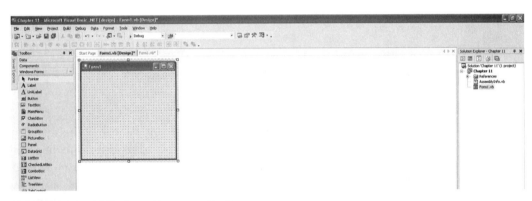

FIGURE 11.1   A Windows Forms application.

PROJECT OVERVIEW

We begin this chapter by looking at the creation of a user interface for our application. The user interface is probably the first step you'll take when developing most typical VB .NET applications. Start the Visual Basic IDE and select Windows Forms Application from the templates that appear.

The next step is to place controls on the default form that appears. There are two separate approaches you can use to do this. First, you can simply double-click on one of the intrinsic Visual Basic controls that appear in the toolbar, which places a single instance of the control on the form. Another way you can place controls on the form begins by clicking the tool in the Toolbox. You then move the mouse pointer to the form window, and the cursor changes to a crosshair. Place the crosshair at the upper-left corner of where you want the control to be, press the left mouse button, and hold it down while dragging the cursor toward the lower-right corner. As you can see in Figure 11.2, when you release the mouse button, the control is drawn.

FIGURE 11.2   You can place controls on a form in several different ways.

You don't have to place controls precisely where you want them because you can move them; Visual Basic provides the necessary tool to reposition them at any time during the development process. To move a control you have created with either process, you click the object in the form window and drag it, releasing the mouse button when you have it in the correct location. You can resize a control very easily as well, by clicking the object so that it is in focus and the sizing handles appear. These handles, which can be seen in Figure 11.3, can then be clicked and dragged to resize the object.

For a first project, you can begin by placing a text box and a command button on the form and position them so that they look similar to Figure 11.4.

The command button needs its Text property changed in the Properties window. It can be changed to "Click." See Figure 11.5.

The next step is to double-click the command button, which brings up the Code window and leaves you something that looks similar to Figure 11.6.

FIGURE 11.3   Handles are useful for positioning and resizing objects.

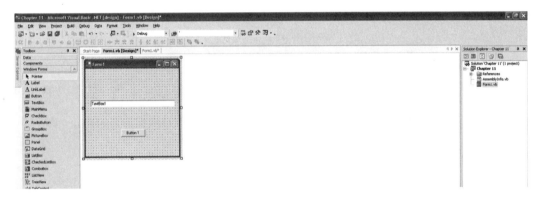

FIGURE 11.4   Beginnings of a GUI.

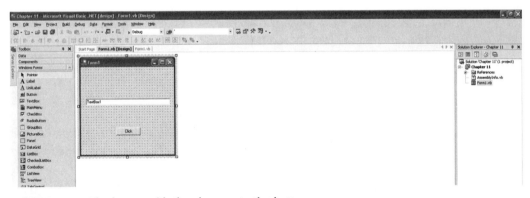

FIGURE 11.5   The layout with the changes to the button.

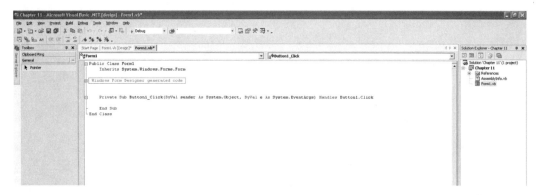

FIGURE 11.6 When you double-click an object on a Visual Basic form, it opens an event in the Code Editor.

Your cursor should be flashing beneath the `Private Sub Command1_Click()` line. Type the following lines into your application:

```
    Private Sub Button1_Click(ByVal sender As System.Object, ByVal e As
System.EventArgs) Handles Button1.Click
Dim strInfo As String
strInfo = "My First Windows Application"
MessageBox.Show("Hello World!")
    End Sub
```

Click on the tab at the top of the screen that says Form1.vb [Design]. Double-click the form to open the `Form_Load` event. Enter the following code and continue reading for an explanation:

```
Private Sub Form1_Load(ByVal sender As System.Object, ByVal e As
System.EventArgs) Handles MyBase.Load

    TextBox1.Text = "Form Load"

End Sub
```

ON THE CD

The CD-ROM that is included with the book contains all of the sample code for each of the projects we create throughout the book. This saves you time and programming mistakes, which allows you to focus only on the task at hand—learning VB .NET. It also contains several additional applications. Please see the CD-ROM for a complete list of projects and applications.

THE CODE EXPLANATION

That's all we need for this application. Although this project is very simple, we did use variables for storing text information in order to point out several features of VB .NET. First, there are only a few simple rules you should keep in mind when using variables. They should be less than 40 characters, they can include letters, numbers, and underscores (_), they cannot use one of the Visual Basic reserved words (i.e., you cannot name a variable Text), and they have to begin with a letter. Let's look carefully at what the code does.

The line that begins with `Private Sub Button1_Click` tells Visual Basic to run this line of code when someone clicks the command button you created called Button1. When they do, the lines of code you typed in are executed. The `End Sub` simply informs Visual Basic that it's time to stop running the code. This event and subsequent code was created automatically for you when you double-clicked on Button1 in a previous step. The `Form_Load` event was created automatically for you when you double-clicked the form. Inside the event, you added a line that sets the text box equal to "Hello World!"

Similar to the `Button1_Click` event, the code is run only when the form is loaded when the program runs. The first line of code that you entered dimensions the variable `strInfo` as a string. The next line assigns the text string "Form Load" to the variable and the final line uses the `MessageBox.Show` method to display the message box.

You may have noticed some additional code in your Code Editor that we haven't looked at or mentioned in this example. Specifically, you will see something that says "Windows Form Designer Generated Code" that contains a "+" sign next to it. If you click the "+", you will see an enormous amount of code that was created for you automatically. This code is the underlying code that is created when you create a form. We'll ignore this section of code for the examples in this book. Because the code is created by VB .NET automatically, we don't really have a need to list it.

RUNNING THE PROGRAM

You can execute the program from within Visual Basic by pressing Ctrl+F5. You should see a window that appears similar to Figure 11.7.

You can close it like any Windows-based application or select the Stop button from within the Visual Basic IDE. You've created your first Windows Forms application. You can save your changes, if you want, by selecting Save from the File

FIGURE 11.7   Your first program
running inside the IDE.

menu. When saving a project, it's best to create a new directory in which you can
store all the files necessary for the project. In this way, you keep the files in one easy
to manage area without the risk of another project corrupting the source code or
data.

COMPLETE CODE LISTING

The following code is the complete listing for this chapter:

```
Public Class Form1
    Inherits System.Windows.Forms.Form

    Private Sub Button1_Click(ByVal sender As System.Object, ByVal e As
System.EventArgs) Handles Button1.Click
        Dim strInfo As String
        strInfo = "My First Windows Application"
        TextBox1.Text = strInfo
        MessageBox.Show("Hello World!")
    End Sub

    Private Sub Form1_Load(ByVal sender As System.Object, ByVal e As
System.EventArgs) Handles MyBase.Load
        TextBox1.Text = "Form Load"
    End Sub
End Class
```

SUMMARY

During the first chapter, we looked at numerous concepts, many of which might be new. We built our first Windows Forms application after dealing with Console applications in the previous chapters. We developed a simple application by using a few intrinsic controls and some basic code, and then proceeded to run it inside the Visual Basic IDE.

Now that you have some of the basics out of the way, let's move to the next chapter—where the real fun begins!

12 Obtaining the Tablet PC SDK

In the preceding chapters, we have looked at the basics of VB .NET development, but up to this point in time, we have yet to get into the specifics of the Tablet PC. In this chapter, we change that as we install the Windows XP Tablet PC Edition Platform Software Development Kit (SDK). In the remaining chapters, we actually look at the various aspects of the development tool as we create a Tablet PC-specific application in each of them.

THE SDK

The Tablet PC Platform SDK has undergone a series of betas since its initial release in 2001. Currently at version 1.5, the Tablet PC Platform SDK is available from the Microsoft Developer Network (MSDN) Web site. At this time, the SDK is specifically available at *http://msdn.microsoft.com/library/default.asp?url=/downloads/list/windevtpc.asp*. You should download it before continuing on with this chapter. The Tablet PC Platform SDK is relatively small and is freely available. The SDK includes several key concepts:

Managed application programming interfaces (APIs): If you remember way back to Chapter 1, The Tablet PC, we talked about the pen and ink capabilities of the Tablet PC. The SDK provides a set of managed APIs for .NET applications that expose these pen and ink features.

Ink controls: Another way to incorporate ink into an application is the ActiveX and .NET controls. The controls include InkPicture and InkEdit. The controls are very quick and easy to use, but unlike the managed APIs, the Ink controls expose a subset of the available underlying functionality. The InkEdit control is the more restricted of the two.

Component Object Model (COM) APIs: If you are interested in developing outside of .NET, COM APIs are available for C++ and Visual Basic 6.

Runtime libraries: To make the development of Tablet PC applications possible on non-Tablet PC editions of Windows, the SDK installs the necessary runtime libraries.

We look at these areas in more detail later in the chapter.

INSTALLING THE TABLET PC PLATFORM SDK

In this section, we provide an overview of how to install the Tablet PC Platform SDK. If you have already installed the SDK and have verified its installation by compiling a few of the samples that are included with it, you can skip this section entirely. If you take this approach and later realize you have problems, you can always refer back to this section of the chapter.

System Requirements

The Tablet PC SDK does not have a great deal of specific hardware or software requirements. Of the requirements, there are some variations that may make your development easier (more on this later). The following operating systems and development environments are required, whereas the digitizer hardware is optional:

Operating system: Your development machine should be running a minimum of Windows 2000 or Windows XP; it is preferable to have the Tablet PC Edition of Windows XP. If you are using an actual Tablet PC, you have the correct OS. Windows 2000 will also work if you do not have Windows XP or Windows XP Tablet PC Edition.

Development environment: You can use either Visual Studio 6 for unmanaged development or Visual Studio .NET for either unmanaged or managed development.

Pointing device: Although a digitizer pad is not required for development, it is recommended. Trying to test input with a mouse is very difficult under the best of circumstances. Of course, if you are using a Tablet PC, you already have a Human Interface Device (HID) to use.

Hard drive: You will need at least 50MB of free hard drive space to install the SDK.

Although the SDK was created to develop applications for Tablet PCs, it can be installed and used on any PC running Windows 2000 and Windows XP Professional or Home versions. This is a nice feature for those without a Tablet PC because it allows for development and testing of Tablet PC software on any machine to which you have access. Unfortunately, development outside of an actual Tablet PC comes with a price. There are a few limitations you should keep in mind for developing without a Tablet PC:

- If you develop an application that relies on Ink Recognition, it will work only on computers running Windows XP Tablet PC Edition.
- Ink controls are limited outside of a Tablet PC, where new ink cannot be captured and existing ink cannot be recognized.
- You will probably want to have access to a digitizing pad if you are using a desktop PC to test pen entry.

With the previously listed problems in mind, Tablet PC application development is best done on a Tablet PC. Even this is not a perfect option because developing on a Tablet PC is a bit cumbersome when you are accustomed to a desktop PC. You could also use the Tablet PC only for remote debugging, in which you create your code on a desktop PC and then test it on a Tablet PC. None of the options are perfect, but any one or combination of these options can be used.

INSTALLING THE SDK

We'll assume that you have already downloaded the Tablet PC SDK. The downloaded file is probably titled "Tablet PC Platform SDK v1.5.setup.exe." Browse to the directory where you downloaded the file and run the setup file. There is nothing special about the SDK installation, but there are some steps to getting your development environment set up correctly.

When you start the installation, you are presented with an option to pick the items to install. You will probably want to install the samples of the SDK because they provide quick examples to many of the SDK features. In fact, if you have the space, it is advisable to install all of the optional components, as shown in Figure 12.1. These optional components also include the distributable merge module (Mstpcrt.msm) that can be redistributed with your applications. You only need this if you are building an application that needs to run on something besides a Tablet PC, such as a standard PC running Windows 2000 or Windows XP.

The rest of the setup is self-explanatory so we'll move on to setting up your development environment to utilize the SDK.

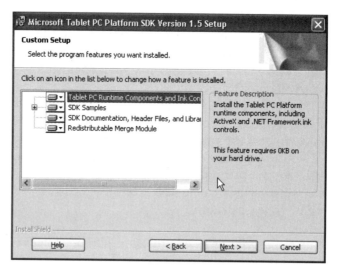

FIGURE 12.1 The optional components.

 We'll assume that you install the SDK at the default path of C:\Program Files\Microsoft Tablet PC Platform SDK. You can choose any path you want, but we assume you are installing it to this path for the purposes of this text. Please remember the location of your path if you install it elsewhere.

Setting up Visual Studio .NET

To build Tablet PC applications using either Visual C# or Microsoft Visual Basic .NET, you must add a reference to Microsoft Tablet PC API to your project in Visual Studio .NET (or use the InkEdit and InkPicture controls). This provides access to the Tablet PC managed object model. To add a reference to the SDK in Visual Studio .NET, perform the following steps:

1. On the Project menu, click Add Reference (see Figure 12.2).
2. On the .NET tab in the Add Reference dialog box, in the Component Name list, select Microsoft Ink Resources, version 1.0.2201.2. If you want to use the Divider or PenInputPanel objects in your application (you'll use these extensively later in the book but if you want, you can browse the SDK documention for additional information), you should also select Microsoft Ink Resources version 1.5.3023.0 (see Figure 12.3). These are commonly shortened to only the component names Microsoft Ink Resources and Microsoft Ink Resources 1.5.
3. Click Select, and then click OK.

FIGURE 12.2   Choose Add Reference from the Project menu.

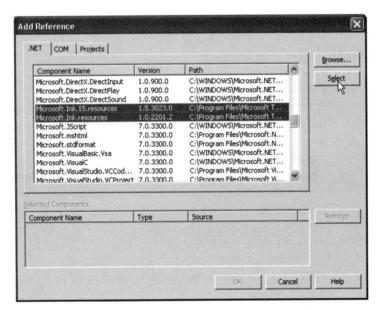

FIGURE 12.3   You need to select the Tablet PC Components.

ActiveX and Managed Controls

Microsoft was kind enough to include ActiveX counterparts for the managed InkEdit and InkPicture controls. These are perfect for users of the ever popular VB6. To build Tablet PC applications in Microsoft Visual Basic 6, you need to reference the Microsoft Tablet PC Type Library, version 1.0 (InkObj.dll), which is installed to the C:\Program Files\Common Files\Microsoft Shared\Ink folder by default. We look at the InkPicture and InkEdit managed components in Chapter 13, Introduction to Ink-Enabled Applications, as we use them to build applications.

There are a few additional things you need to do in order to use the control. First, you need to add the InkConstants.bas file, found in the same folder as the InkObj.dll file to your project. Now, to use the InkEdit ActiveX control in your application, you should right-click the Toolbox and select Components. Add a reference to the Microsoft InkEdit control version 1.0 (InkEd.dll), which should be located in the C:\Windows\System32 folder. Lastly, to use the InkDivider object, you need to create a reference in your project to the InkDiv.dll found in C:\Program Files\Common Files\Microsoft Shared\Ink folder by default.

Microsoft Visual C++

To build Tablet PC applications in Microsoft Visual C++®, you will need to update the system environment variables, set up directory options for Visual Studio, and access the Tablet PC interfaces in your project. Follow these steps to build Tablet PC programs:

1. Click the Start button and then click Control Panel.
2. If you are in the Classic View (see Figure 12.4), you need to double-click the System icon. On the other hand, if you are in Category View (see Figure 12.5), you should click the Performance and Maintenance icon first and then click the System icon.
3. Choose the Advanced tab and then click the Environment Variables button. This displays an Environment Variables dialog box (see Figure 12.6).
4. Under System variables in the Environment Variables dialog box, select Path, and then click Edit.
5. In the Edit System Variable dialog box, add ";%CommonProgram-Files%\Microsoft Shared\Ink" to the Variable value, and then click OK. This can go at the end of anything already existing and should be entered without the quotes.
6. Under System variables in the Environment Variables dialog box, select INCLUDE, and then click Edit.
7. In the Edit System Variable dialog box, add "; %ProgramFiles%\Microsoft Tablet PC Platform SDK\Include" to the Variable value, and then click OK. If you installed the SDK to another folder, you may need to alter the paths in the previous steps.

8. Click OK in the Environment Variables dialog box.
9. Click OK in the System Properties dialog box.
10. In Visual Studio, select the Tools menu and then click Options. In Visual Studio .NET, the Visual C++ Directories options are located under the Projects node.
11. On the Directories tab, in the Show directories for list, select Include files.
12. Under Directories (the Include Directories list in Visual Studio .NET), add "%ProgramFiles%\Microsoft Tablet PC Platform SDK\Include," and then press Enter.
13. In the Show directories for list, select Library files.
14. Under Directories (the Library Directories list in Visual Studio .NET), add "%CommonProgramFiles%\Microsoft Shared\Ink," and then press Enter.
15. Click OK.

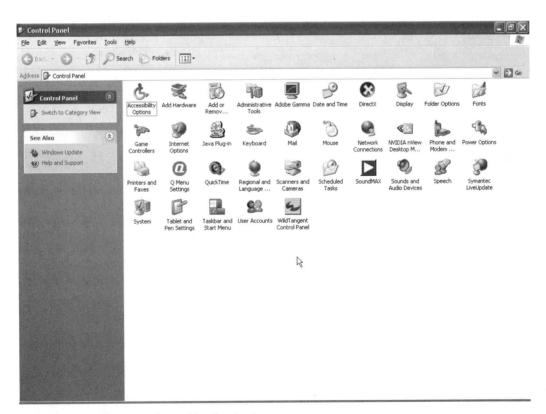

FIGURE 12.4 The Control Panel in Classic View.

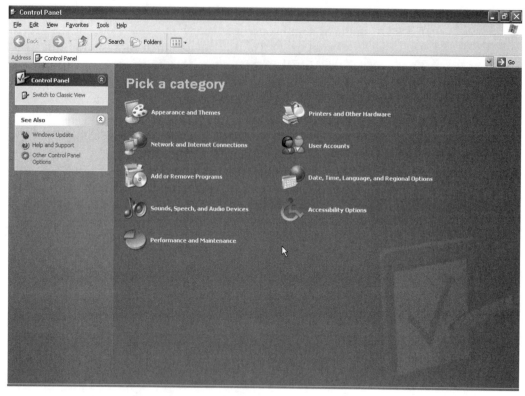

FIGURE 12.5   The Control Panel in Category View.

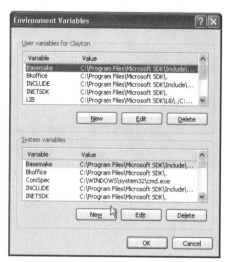

FIGURE 12.6   The Environment Variables dialog box.

To access the Tablet PC interfaces, you must include the Msinkaut.h and Msinkaut_i.c files and to access the InkEdit control interfaces, you need to include Inked.h and Inked_i.c. in your project like the following:

```
#include <msinkaut.h>
#include <msinkaut_i.c>
#include <inked.h>
#include <inked_i.c>
```

You could use the following import directive instead of the first two #include *statements:* "#import "InkObj.dll" no_namespace exclude("tagXFORM")".

The files we have been using are only a few of the many that are installed with the Tablet PC SDK. According to Microsoft documentation, there are over 160 files included with the SDK. We are not reviewing each of the files, but the following list details some of the more important ones:

Microsoft.Ink: Most of the managed API is implemented within this assembly.

InkObj.dll: This includes most of the core ink services used internally by the platform and includes the COM automation APIs and type libraries.

Tpcsdk10.chm: The help documentation for the SDK is available here.

MSInkAut.h: The main COM automation API header is available here to include in your C++ projects.

InkEd.dll: The InkEdit control is implemented in this dynamic-link library (DLL).

Wisptis.exe: This runs as a system service that provides pen-data collection for other components of the SDK. When a component needs to interact with the pen, this executable is spawned as a service to communicate directly with the input device. This interacts with the digitizer on a Tablet PC or a mouse on a desktop PC.

USING THE SDK

Most of the applications developed in this book utilize VB .NET, although we also use C# and C++. We look at some of the most important aspects of the SDK before we begin writing our first Tablet PC application.

Managed APIs

If you remember back to Chapter 4, Basics of the .NET Framework, a significant portion of the Microsoft .NET Framework is the common language runtime (CLR), which controls and supports the execution of .NET applications written in various languages. With this in mind, the Tablet PC Platform SDK APIs are "managed" as they were created to run in the .NET Framework's common language runtime. As such, you can call the APIs from any .NET language, such as C#, C++, Jscript, or VB .NET.

The managed APIs are divided into three distinct areas that encompass the functions required for a Tablet PC:

The Tablet Input API (Pen API): As the name suggests, the Pen API is targeted at pen-specific features, such as the various buttons on a pen. It also collects digital ink and gestures from the movement of the pen.

The Ink Data Management API (Ink API): Once ink has been collected, the Ink API takes over for its manipulation and storage.

The Ink Recognition API (Recognition API): The Recognition API is used to recognize the ink.

INK CONTROLS

There are two Ink controls that are included in the Tablet PC Platform SDK. They allow us to quickly integrate pen and ink functions into a new or existing application. The controls can be easily dragged onto forms in a language that supports forms-based software development, such as VB. If you remember the section earlier in this chapter, there are ActiveX Ink controls and managed Ink controls. We use these in Chapter 13, Introduction to Ink-Enabled Applications.

The following controls are available for us to use:

InkEdit: This control is an extension of the common RichEdit control and captures and can also convert ink into text. This is the quickest way to add conversion to an application.

InkPicture: This control is used to display or capture ink over existing images.

TABLET PC SPEECH

When dealing with Tablet PC development, it is quite obvious that we need to take advantage of ink for data entry. But, it is definitely not the only one. As the Tablet PC was created with portability in mind in which a keyboard is not always readily available, speech also plays an important role in software development. We are going to delve into the basics of the Microsoft Speech API (SAPI) SDK beginning in Chapter 21, Speech Input with SAPI. Along the same lines, we are also going to take advantage of Microsoft Agent (Chapters 19 and 20, Getting Started with Microsoft Agent and Advanced Microsoft Agent, respectively) technologies for a couple of example projects. We go over more of each of them as we encounter them.

SUMMARY

In this chapter, we have installed the Tablet PC SDK and taken a quick glance at some of its features. In the remaining chapters, we actually look at the more involved aspects of the development tool as we create a Tablet PC-specific application in each of them.

13 Introduction to Ink-Enabled Applications

In this chapter, we build our first ink-enabled applications using VB .NET. We also use source code for C# in this chapter so that you can see the differences and similarities between the languages. For the remaining chapters, we provide source code in VB .NET, but if you prefer C#, it won't take you very long to convert the samples.

ON THE CD

The source code for the projects are located on the CD-ROM in the PROJECTS folder. You can either type them in as you go or you can copy the projects from the CD-ROM to your hard drive for editing.

INTRODUCTION TO INK CONTROLS

In Chapter 12, Obtaining the Tablet PC SDK, we looked at the InkEdit and InkPicture controls. Now, in this chapter, we learn how to utilize the controls to build fully aware ink applications.

The InkPicture and InkEdit controls allow a developer to quickly add ink capabilities to an application. Depending on the needs for a given project, a developer may need the ability to have ink conversion to text, ink for annotating an existing image, or countless other variations. These two controls provide enough functionality to take care of most of the projects you will develop.

Differences between the Controls

The InkEdit control was derived from the standard RichTextBox class and is the control of choice when you want to perform handwriting recognition. By default, the control recognizes the text as you enter it and automatically displays it within the text display area. If you are using a keyboard, you can also enter text into the control, which provides most of the functionality of the RichTextBox. In addition to collecting ink and recognizing and displaying it in text form, the InkEdit control also allows you to display ink as an embedded object, with text formatting options such as underline.

Like the InkEdit control, the InkPicture control is also derived from a standard control. This time, it's the Picture class and it offers an area on which you can annotate. This control is most often used when you do not have a need for ink recognition. Popular uses include capturing signatures, or because it includes the ability to display an image that the user can draw on, it can be used for marking up a scanned form or even drawing. The control accepts a variety of formats, such as jpg, bmp, gif, or png.

USING THE INKEDIT CONTROL

As was mentioned previously, the InkEdit control was derived from the RichTextBox class. In its default state, the control automatically captures ink, recognizes the handwriting, "erases" the ink, and then displays the recognized text inside the text box. Interestingly, the handwriting is not really erased, and you can alter between the recognized text and the handwritten information at runtime. Recognition occurs very quickly and allows the user to enter relatively large amounts of information.

Creating a Project

To create a project with the InkEdit control, you need to create a new Windows Forms application and then add it to your Toolbox. Follow these steps to add the control to the toolbox:

1. Create a new project.
2. Right-click on the Toolbox.
3. Select Customize.
4. Select the .NET Framework Components tab.
5. Check InkEdit and then click OK.

 There is another way that you can access these controls. Instead of adding them to your Toolbox, you can reference and create it in code. However, it's much easier to add them to the Toolbox.

After you have the InkEdit control in the Toolbox, you can add it to your form as quickly as you can to any standard application. To add an instance of the InkEdit control to your form, select the InkEdit control in your Toolbox, and then draw a rectangle region on your form. The size you draw the control is very important because the inkable region is located within the control itself. As with much of Tablet PC development, user interface design is important. If the control is too small, it is difficult for the user to write within its boundaries and the resulting recognition will not be very good. At a minimum, you should size the control to allow a few words and preferably more.

As a reminder, the InkEdit control automatically collects ink and gestures (a sort of shorthand that we discuss in greater detail in Chapter 18, Using Gestures to Control Tablet Media Player). This can be a problem if you are developing only on a desktop PC because you do not have the ability to draw within the control. You can use your mouse for input by adding a single line of code:

```
InkEdit1.UseMouseForInput = True
```

Of course, if you are developing on a Tablet PC, this will not be an issue. Either way, after you have added the control to a form, you can run the application in the IDE. Once started, you can write on the InkEdit control. As you can see, the control already provides the ability to write on it without a single line of code (unless you set it up to handle mouse input, which is discussed above) and also converts ink to text automatically. Figures 13.1 and 13.2 display an example of ink and the resulting text.

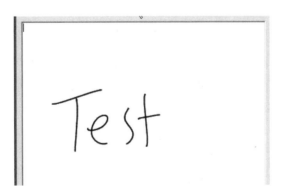

FIGURE 13.1   Ink being entered.

You can change many aspects of how the InkEdit control recognizes ink. For example, you have the ability to alter the size and position of ink into the control. By setting various properties, you could quickly alter the control to capture a sig-

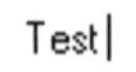

Test|

FIGURE 13.2 Conversion
has taken place.

nature or write a handwritten letter. The properties are quite simple, and are discussed in the following sections.

Property Description

This section details the properties available for InkEdit.

Active X Only

We'll look at the properties that are specific to the Active X control first.

Appearance: Returns or sets a value that determines whether the control appears flat or 3D

BackColor: Returns or sets the background color for the control

BorderStyle: Returns or sets a value that determines whether the control has a border

DisableNoScroll: Returns or sets a value that determines whether scroll bars in the control are disabled

DragIcon: Returns or sets the icon that appears as the pointer in a drag-and-drop operation

Font: Returns or sets the font of the text that the control displays

HWnd: Returns the window handle to which the control is bound

Locked: Returns or sets a value that specifies whether the control is read-only

MaxLength: Returns or sets a value indicating whether an InkEdit control can hold a maximum number of characters and, if so, specifies the maximum number of characters

MouseIcon: Returns or sets the current custom mouse icon

MousePointer: Returns or sets a value that indicates the type of mouse pointer that appears when the mouse is over a particular part of the control

MultiLine: Returns or sets a value that indicates whether this is a multiline control

ScrollBars: Returns or sets the type of scroll bars that appear in the control

SelAlignment: Returns or sets the alignment to apply to the current selection or insertion point

SelBold: Returns or sets a value that specifies whether the font style of the currently selected text in the control is bold

SelCharOffset: Returns or sets whether text in the control appears on the baseline, as a superscript, or as a subscript

SelColor: Returns or sets the text color of the current text selection or insertion point

SelFontName: Returns or sets the font name of the selected text within the control

SelFontSize: Returns or sets the font size of the selected text within the control

SelItalic: Returns or sets a value that specifies whether the font style of the currently selected text in the control is italic

SelLength: Returns or sets the number of characters that are selected in the control

SelRTF: Returns or sets the currently selected Rich Text Format (RTF) formatted text in the control

SelStart: Returns or sets the starting point of the text that is selected in the text box

SelText: Returns or sets the selected text within the control

SelUnderline: Returns or sets a value that specifies whether the font style of the currently selected text in the control is underlined

Text: Returns or sets the current text in the text box

TextRTF: Returns or sets the text of the control, including all Rich Text Format (RTF) codes

Managed Library Only

The Managed Library has its own set of properties.

CreateParams: Returns the required creation parameters when the control handle is created

Cursor: Returns or sets the cursor that appears when the mouse pointer is over the control

DrawingAttributes: Returns or sets the drawing attributes to apply to ink as it is drawn

This property behaves differently than the DefaultDrawingAttributes *property of the* InkCollector *object, the* InkOverlay *object, and the InkPicture control. Whereas the* DefaultDrawingAttributes *property specifies the drawing attributes that are applied to a new cursor, the* DrawingAttributes *property specifies the drawing attributes for ink that is yet to be collected by the InkEdit control.*

Enabled: Set to True or False to enable or disable

General

There are properties that are available for both the Active X and Managed Libraries.

Factoid: Returns or sets the factoid that a recognizer uses to constrain its search for the recognition result

InkInsertMode: Returns or sets the value that specifies how ink is collected when drawn on the control

InkMode: Returns or sets a value that specifies whether ink collection is disabled, ink is collected, or ink and gestures are collected

RecoTimeout: Returns or sets the number of milliseconds after an ink stroke has ended that text recognition begins

Recognizer: Returns or sets the recognizer to use for recognition

SelInks: Returns or sets the array of embedded Ink objects (if displayed as ink) that the current selection contains

SelInksDisplayMode: Returns or sets a value that allows toggling the appearance of the selection between ink and text

Status: Returns a value that specifies whether the control is idle, collecting ink, or recognizing ink

UseMouseForInput: Returns or sets a value that indicates whether the mouse is used as an input device

Having Fun with Properties

With so many properties that we have access to, there is a great deal of fun we can have with them. Before we go further, we need to add the following line to the top of the project:

```
Imports Microsoft.Ink
```

Next, we are going to add a series of buttons that will allow us to set some of the various properties for the control. Add the following buttons with their respective properties (see Table 13.1) and place them on the form using Figure 13.3 as an example.

TABLE 13.1 Adding buttons and properties to our example

Name	Text
btnDisplayAsInk	Display as Ink
btnDisplayAsText	Display as Text
btnInsertAsInk	Insert as Ink
btnInsertAsText	Insert as Text
btnIncreaseRecoTime	+ Reco Time
btnSubRecoTime	– Reco Time

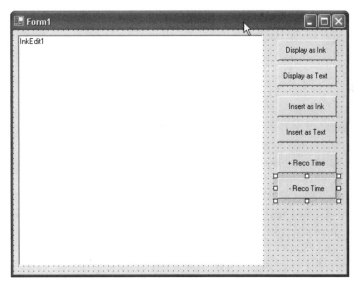

FIGURE 13.3 Button layout for our example.

You can probably tell what each of these controls will be used for by their names and text properties. We'll begin with btnDisplayAsInk. Double-click the control to open the Code Editor and create the `btnDisplay_Click` event procedure. We need to add two lines to the event:

```
InkEdit1.SelectAll()
InkEdit1.SelInksDisplayMode = InkDisplayMode.Ink
```

The previous code begins by selecting all of the text located with `InkEdit1`, which was added earlier in the chapter. The text needs to be selected before changing its display mode to ink, which is accomplished in the second line. If you do not first select the ink, you will not see a change in your control. It's worth noting that the InkEdit control does not try to convert text, which was entered into the control as text, into ink. Obviously, it would be impossible for the control to try to simulate what your handwriting should look like so that it could convert it into a mock form of writing.

We'll continue the process by adding the opposite code—displaying ink as text. We can begin by choosing "btnDisplayAsText" from the class name listbox and then choose "click" from the method name listbox. This creates the click event for the control for us. We need to add the following code to the event:

```
InkEdit1.SelectAll()
InkEdit1.SelInksDisplayMode = InkDisplayMode.Text
```

This code, like most in this section of the chapter, works similarly to the previous code. Again, it begins by selecting the text and then changes the display mode, but this time, it converts it to text.

Not only can we change the way ink is displayed, but we can also determine in what format the control actually collects ink. Remember that if you set this to insert the ink as text, it actually converts the ink to text, and, therefore, you cannot get the ink back. On the other hand, if you insert as ink, you can display it as ink or text and change it back and forth as many times as you want because there is no conversion actually taking place. Additionally, when the control tries to convert ink to text, it also tries to determine what type of word was entered and, therefore, may make changes you were not intending. These properties play a crucial role in the way the ink is utilized, so depending on the type of application you are trying to create, the correct property is extremely important.

To change the way the ink is inserted, you need a single line of code entered into the `btnInsertAsInk` and `btnInsertAsText` click events (refer back earlier in the chapter for information on creating the events):

```
Private Sub btnInsertAsInk_Click(ByVal sender As System.Object, ByVal e
As System.EventArgs) Handles btnInsertAsInk.Click
```

```
        InkEdit1.InkInsertMode = InkInsertMode.InsertAsInk
End Sub

Private Sub btnInsertAsText_Click(ByVal sender As System.Object, ByVal
e As System.EventArgs) Handles btnInsertAsText.Click
        InkEdit1.InkInsertMode = InkInsertMode.InsertAsText
End Sub
```

The final property we are going to work with at this time is the `RecoTimeout` property, which determines how long it takes for the control to begin the recognition process after your pen leaves the control. The property is set using milliseconds. You can create the `btnSubRecoTime` and `btnIncreaseRecoTime` events and enter the following code:

```
Private Sub btnSubRecoTime_Click(ByVal sender As System.Object, ByVal e
As System.EventArgs) Handles btnSubRecoTime.Click
    If InkEdit1.RecoTimeout > 100 Then
        InkEdit1.RecoTimeout = InkEdit1.RecoTimeout - 100
    End If
End Sub

Private Sub btnIncreaseRecoTime_Click(ByVal sender As System.Object,
ByVal e As System.EventArgs) Handles btnIncreaseRecoTime.Click
    If InkEdit1.RecoTimeout < 10000 Then
        InkEdit1.RecoTimeout = InkEdit1.RecoTimeout + 100
    End If
End Sub
```

The code uses a simple `If...Then` statement to determine how small the current timeout values are set. Its purpose is to limit how quickly the recognition occurs because a negative value would obviously cause problems. You can now save your project and run the application. You can try the various properties to see how they work. Figure 13.4 displays an example of what you will see as you set properties. For additional information, the default value for `RecoTimeout` is 2 seconds or 2000 milliseconds. Changing this value, like the earlier values, can play an important role in your applications. If you are writing some general purpose applications, 2 seconds is probably a good starting point. However, if you are writing something, such as a game, that recognizes what you are writing, you may need to shorten the value, or if you are creating an application for children, you may want to lengthen the time to allow for their generally slow handwriting speed.

There are many other properties we could work with, but they are all covered extensively in the SDK. In addition, we also mention the specific properties we use during the creation of our many applications as we progress through the book.

FIGURE 13.4 Setting various properties changes the way ink is displayed.

Saving Content

After you have some content entered into the control, it is very easy to save it to disk and then reload it when needed. You simply need to call the SaveFile method and the LoadFile method to save to a file or to a stream. As an example, let's extend the application we have been working on to save the contents to "C:\myinktext.rtf."

First, add buttons for loading and saving to the form with the properties shown in Table 13.2.

TABLE 13.2 Adding buttons for loading and saving to the form

Name	*Text*
btnSaveFile	Save File
btnLoadFile	LoadFile

The next step is to add a single line of code to the btnSaveFile and btnLoadFile click events. The line calls the SaveFile method and saves the file to C:\MyInk-Test.rtf. You will probably notice the RTF extension on the file that is given as the InkEdit control is based on the RichTextBox.

First, here is the code for the `btnSaveFile` event procedure:

```
InkEdit1.SaveFile("C:\MyInkTest.rtf")
```

To load the file, you should call the `LoadFile` method. You can open the file with the following line of code to the `btnLoadFile` click event procedure:

```
InkEdit1.LoadFile("C:\MyInkTest.rtf")
```

The sample you have now allows you to set various properties of the InkEdit control. You can also load and save files albeit in an extremely simple manner. There is one final thing that you can add to the program. To erase the text in `InkEdit1` that is visible on startup, you can add the following line to the `Form_Load` event:

```
InkEdit1.Text = ""
```

That's it for this simple example. You can now save it and then create a new project as we now turn our attention to the InkPicture control.

USING THE INKPICTURE CONTROL

If you have not already done so, you need to create a new Windows Forms application and then add the InkPicture control to the form:

1. Right-click on the Toolbox.
2. Select Customize.
3. Select the .NET Framework Components tab.
4. Check InkPicture and then click OK.

You can now add an instance of the InkPicture control to your form as it is now visible in the Toolbox. You can select it and then draw a rectangle shape on your form. It will be positioned in the area in which you draw the rectangle.

The InkPicture control has some unique properties that we have the ability to manipulate. By default, the control captures ink and displays the ink so you don't have to do anything special with it. If you were to save and then run the current sample, you would see that the control does allow you to write with your pen (see Figure 13.5).

You can close the running sample to return to the IDE, and we'll look at some of the properties provided by the InkPicture control.

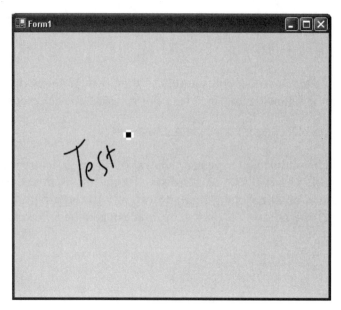

FIGURE 13.5 The control displays ink by default.

Property Description

Like InkEdit, InkPicture also has its own set of properties. We'll look at the ActiveX properties, the Managed Library properties, and then a set of properties that are general to both.

ActiveX Only

The ActiveX control provides a set of properties that are specific to it.

hWnd: Returns the window handle to which the control is bound

MouseIcon: Returns or sets the current custom mouse icon

MousePointer: Returns or sets a value that indicates the type of mouse pointer that appears when the mouse is over a particular part of the control

tbpropicturePicture: Returns the graphics file to appear on the control

Managed Library Only

The Managed Library has a set of properties that are unique.

AccessibleDescription: Returns or sets the description of the control that the accessibility client applications use

AccessibleName: Returns or sets the name of the control that the accessibility client applications use

AccessibleRole: Returns or sets the role of the control that the accessibility client applications use

Anchor: Returns or sets which edges of the control are anchored to the edges of its container

BackgroundImage: Returns or sets the background image that appears in the control; allows you to add a picture that can be used for markup

BorderStyle: Returns or sets the border style for the control

ClipInkToMargin: Returns a value that specifies whether to clip strokes when they are outside the default margin

CollectingInk: Returns the value that specifies whether the control is collecting ink

ContextMenu: Returns or sets the shortcut menu to appear when the user right-clicks the control

DefaultMargin: Returns the default margin that the MarginX and MarginY properties use

Dock: Returns or sets to which edge of the parent container the control is docked

Handle: Gets the window handle to which the control is bound

Image: Returns or sets the image that appears in the control

Location: Returns or sets the coordinates of the upper-left corner of the control relative to the upper-left corner of its container

Size: Returns or sets the height and width of the control in pixels

General

The ActiveX and Managed Libraries both offer these properties.

AutoRedraw: Returns or sets a value that specifies whether the InkPicture repaints when the window is invalidated

BackColor: Returns or sets the background color for the control. The default background color is the system window background color, which is typically white

CollectionMode: Returns or sets the collection mode that determines whether ink, gestures, or ink and gestures are recognized as the user writes

Cursor: Returns or sets the cursor that appears when the mouse pointer is over the control

Cursors: Returns the number of cursors available for use in the ink-enabled region; each cursor corresponds to the tip of a pen or other ink input device

DefaultDrawingAttributes: Returns or sets the default drawing attributes to use when collecting and displaying ink

DesiredPacketDescription: Returns or sets the packet description of the control

DynamicRendering: Returns or sets the value that specifies whether the control dynamically renders the ink as it is collected

EditingMode: Returns or sets a value that specifies whether the control is in ink mode, deletion mode, or selecting/editing mode

Enabled: Returns or sets a value that determines whether the control can respond to user-generated events

EraserMode: Returns or sets the value that specifies whether ink is erased by stroke or by point

EraserWidth: Returns or sets the value that specifies the width of the eraser pen tip

 This property also appears in the InkOverlay *object.*

Ink: Returns or sets the Ink object that is associated with the InkPicture control

InkEnabled: Returns or sets a value that specifies whether the InkPicture control collects pen input

MarginX: Returns or sets the x-axis margin around the window rectangle in screen coordinates

MarginY: Returns or sets the y-axis margin around the window rectangle in screen coordinates

Selection: Returns or sets the collection of ink strokes that are currently selected

SizeMode: Returns or sets how the control handles image placement and sizing

SupportHighContrastInk: Returns a value that specifies whether ink is rendered as just one color when the system is in High Contrast mode

SupportHighContrastSelectionUI: Returns or sets a value that specifies whether all selection user interfaces are drawn in high contrast when the system is in High Contrast mode

Tablet: Returns the Tablet object that the InkPicture control is currently using to collect input; returns the object that represents the tablet hardware, such as manufacturer information

If you quickly browse the list, you will see some interesting properties and many will already look familiar to you because the InkEdit control shares many of the same properties.

As you remember, the InkPicture control is most often used to annotate images. With that in mind, it's important to learn about the properties that control ink selection and ink erasing. The first thing we'll deal with is removing (erasing) ink from the control. Your application can offer two erasing modes:

Point erase: Allows the user to erase a single pixel at a time

Stroke erase: Allows the user to erase the entire stroke, if the user touches anywhere on a stroke

Supporting erase functionality is imperative to many applications because it allows users to easily erase and reenter drawings that they have produced with ink. With the form you already have, add two buttons to the form with the properties shown in Table 13.3.

TABLE 13.3   Adding btnErasePoint and btnEraseStroke to the form

Name	Text
btnErasePoint	Erase Point
btnEraseStroke	Erase Stroke

Your form should now look like Figure 13.6.

It's now very quick and easy to add the erasing capability. Create the click events for both controls and then add the following code into the appropriate procedures:

```
Private Sub btnErasePoint_Click(ByVal sender As System.Object, ByVal e
As System.EventArgs) Handles btnErasePoint.Click
    InkPicture1.EraserMode = InkOverlayEraserMode.PointErase
    InkPicture1.EditingMode = InkOverlayEditingMode.Delete
End Sub

Private Sub btnEraseStroke_Click(ByVal sender As System.Object, ByVal e
As System.EventArgs) Handles btnEraseStroke.Click
    InkPicture1.EraserMode = InkOverlayEraserMode.StrokeErase
    InkPicture1.EditingMode = InkOverlayEditingMode.Delete
End Sub
```

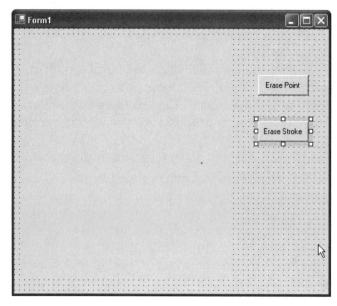

FIGURE 13.6 Your form now has buttons added for ink erasing.

The first line of code in each procedure sets the EraserMode property to Point or Ink respectively, whereas the second line of code sets the EditingMode property to Delete. These are the only two properties required for erasing. You should save the application and then test out the erasing. At this time, you can erase, but you don't have a way to begin drawing again. This can be handled with another button and a single line of code. Add the button shown in Table 13.4 to the form.

TABLE 13.4 Adding btnDraw to the form

Name	Test
btnDraw	Draw

Next, in the button's click event, you can add the following line of code, which again uses the EditingMode property, but now sets it to Ink:

```
InkPicture1.EditingMode = InkOverlayEditingMode.Ink
```

You can now move back and forth between the two different types of ink deletion and can also support drawing. Along these same lines, you might want to give the user the ability to select ink. Again, we can use the `EditingMode` property, but this time we can set it to `Select`. Add another button to the form called btnSelection and change its Text property to "Selection." Within the button's click event procedure, you can add the following line of code:

```
InkPicture1.EditingMode = InkOverlayEditingMode.Select
```

Your application now provides the ability to select ink, erase ink, and draw ink. We're now going to look at how we can alter the appearance of the ink collected in the control beginning with antialiasing. When you typically draw a line on a computer screen, its edges often have jagged edges. To antialias the controls so that they appear smooth, you use the `Form_Load` event and add the following line of code:

```
InkPicture1.DefaultDrawingAttributes.AntiAliased = True
```

Now, as you draw on the InkPicture control, the ink will be antialiased. There are certainly a number of additional properties, but these are some of the most important. In later chapters, we cover additional properties as we need them and you can refer to the SDK for additional information if needed.

OBJECTS

Along with the controls we have been using in this chapter, the SDK also provides several objects that we can use throughout our applications.

InkCollector **Object**

The `InkCollector` object captures ink input that is placed into a known application window. The `InkCollector` object captures the input in real time and then directs it into an `Ink` object. The ink strokes can then be manipulated or sent to a recognizer for recognition.

InkOverlay **Object**

The next object to look at is the `InkOverlay` object, which is a superset of the `InkCollector` object and provides editing support. Both objects, as does the InkPicture control, use common constructs, such as the `Ink` object and the InkDrawingAttributes collection, so that the basic way to change the color of ink is the same everywhere. This enables you to reuse code and makes it much easier for you to remember. InkOverlay is perfect for annotation in which pen size, ink, color, and position are the most important aspects for a program.

InkOverlay differs from InkCollector in several ways:

- It raises events for begin and end stroke, along with ink attribute changes.
- It enables users to select, erase, and resize ink.
- It supports Cut, Copy, and Paste commands.

The process for using InkOverlay and InkCollector are very similar. Let's create a sample application. First, open VB .NET and create a new Windows Forms application. Next, add a label control to the form and resize it so that it takes up much of the form's visible area. Open the Code Editor and then add references to the following:

Microsoft Tablet PC API

The next step is to add a button to the form and just leave it with its given name of Button1. You can also leave the standard Text property of "Button1." Now, back in the Code Editor, we can use the Form_Load event to programmatically set some of these properties and create the InkCollector:

```
Private Sub Form1_Load(ByVal sender As System.Object, ByVal e As
System.EventArgs) Handles MyBase.Load
    Label1.BackColor = Color.White
    Label1.Text = ""
    Button1.Text = "Recognize It"
    theInkCollector = New Microsoft.Ink.InkCollector(Label1.Handle)
    theInkCollector.Handle = Label1.Handle
    theInkCollector.Enabled = True
End Sub
```

The next step is to add the Button1_Click event, which we'll use to display a message box with the recognized text from our strokes. We'll use a simple Try and Catch to make sure we are executing this on a Tablet PC. This is good programming practice, but because we know that the applications we are developing in the majority of the examples are only going to run on a Tablet PC, we may skip this step. Here is the code:

```
Private Sub Button1_Click(ByVal sender As System.Object, ByVal e As
System.EventArgs) Handles Button1.Click
    Dim strokes As Microsoft.Ink.Strokes = theInkCollector.Ink.Strokes
```

```
    Try
        MsgBox(strokes.ToString(), MsgBoxStyle.OKOnly, "VB.NET API")
    Catch
        MsgBox("Not a Tablet PC", MsgBoxStyle.OKOnly, "Error in VB.NET
API")
    End Try
End Sub
```

The last thing we need to do is use the `Dispose` method of the `InkCollector` to prevent memory leaks. We can use the `Form1_Closing` event as a place to handle this:

```
Private Sub Form1_Closing(ByVal sender As Object, ByVal e As
System.ComponentModel.CancelEventArgs) Handles MyBase.Closing
    theInkCollector.Dispose()
End Sub
Selecting Ink
```

You can now save and run the application. Enter some ink into the white area and then click the button that says "Recognize It." This should display the recognized text in a message similar to the one seen in Figure 13.7.

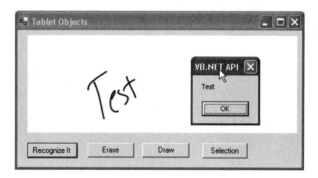

FIGURE 13.7 Our application being tested.

Now, if you want to change the `InkCollector` to an `InkOverlay`, it's as simple as changing these two lines:

```
Dim WithEvents theInkCollector As Microsoft.Ink.InkCollector
theInkCollector = New Microsoft.Ink.InkCollector(Label1.Handle)
```

to read:

```
Dim WithEvents theInkCollector As Microsoft.Ink.InkOverlay
theInkCollector = New Microsoft.Ink.InkOverlay(Label1.Handle)
```

After changing the code, you can check the application to make sure everything works as expected. Now, let's expand upon the application. The `InkOverlay` object enables users to use a Lasso tool to select `Ink` objects or they can select ink by tapping any `Ink` object. It also allows us to erase ink. Like InkPicture, we also need to have a way to get back to Drawing mode. Now, let's add three additional buttons to the application, leaving them as named by default (i.e., Button2, Button3, Button4). We now go into the `Form_Load` event and add the following three lines of code:

```
Button2.Text = "Erase"
Button3.Text = "Draw"
Button4.Text = "Selection"
```

You're already familiar with the properties, so we can simply write the code for each of the events:

```
Private Sub Button2_Click(ByVal sender As System.Object, ByVal e As
System.EventArgs) Handles Button2.Click
    theInkCollector.EraserMode =
Microsoft.Ink.InkOverlayEraserMode.StrokeErase
    theInkCollector.EditingMode =
Microsoft.Ink.InkOverlayEditingMode.Delete
End Sub

Private Sub Button3_Click(ByVal sender As System.Object, ByVal e As
System.EventArgs) Handles Button3.Click
    theInkCollector.EditingMode =
Microsoft.Ink.InkOverlayEditingMode.Ink
End Sub

Private Sub Button4_Click(ByVal sender As System.Object, ByVal e As
System.EventArgs) Handles Button4.Click
    theInkCollector.EditingMode =
Microsoft.Ink.InkOverlayEditingMode.Select
End Sub
```

ADDING INK TO EXISTING APPLICATIONS

The final part of this chapter is devoted to adding ink to existing Windows Forms applications. You can certainly add controls such as InkPicture or InkEdit, or you can use objects such as InkOverlay, but the easiest way to add ink to an existing application is via the PenInputPanel object. As an attachable object, it allows you to add in-place pen input to your applications. You can choose either handwriting or keyboard options for the input method for the Pen Input Panel. Your end user will have the ability to switch between input methods using buttons on the user interface.

Let's create a simple example using the Pen Input Panel. Begin by creating a new Window Forms application in VB .NET. Next, add four standard TextBox controls to the window to simulate an existing VB application. Next, we need to add a reference to the Tablet PC API 1.5. Make sure to select the 1.5 API because version 1 does not include the PenInputPanel object.

The next step is to add the following Imports statement to the Code Editor:

```
Imports Microsoft.Ink
```

Add the following code to create the Pen Input Panel:

```
Dim thePenInputPanel As New PenInputPanel()
```

Now that we have created the Pen Input Panel, we need to attach it to a control before we can use it. We have two options for this. The first is to create three separate Pen Input Panels and then assign each of them to one of the text boxes. The other option is to create the single Pen Input Panel and then attach the control as necessary. In our case, we are going to attach the controls during the Mouse_Down events of the first three TextBox controls. We'll leave the last one alone so we can verify that the Pen Input Panel only works when we attach it to a control.

Here are the procedures and the code:

```
Private Sub TextBox1_MouseDown(ByVal sender As Object, ByVal e As
System.Windows.Forms.MouseEventArgs) Handles TextBox1.MouseDown
    thePenInputPanel.AttachedEditControl = TextBox1
End Sub

Private Sub TextBox2_MouseDown(ByVal sender As Object, ByVal e As
System.Windows.Forms.MouseEventArgs) Handles TextBox2.MouseDown
    thePenInputPanel.AttachedEditControl = TextBox2
End Sub

Private Sub TextBox3_MouseDown(ByVal sender As Object, ByVal e As
System.Windows.Forms.MouseEventArgs) Handles TextBox3.MouseDown
    thePenInputPanel.AttachedEditControl = TextBox3
End Sub
```

You can now test the application. You should try to use your pen to click in the first three text boxes to verify that it works (see Figure 13.8). Additionally, try clicking in the fourth text box to verify that the Pen Input Panel does not appear. Lastly, if you have a mouse attached, try clicking in the text boxes with a mouse. This should verify that the Pen Input Panel does not work unless a pen is being used.

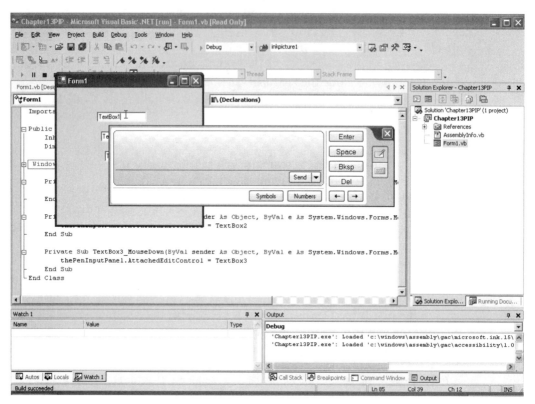

FIGURE 13.8   The Pen Input Panel is displayed.

SUMMARY

In this chapter, we were able to look at the InkEdit and InkPicture controls and also a few of the objects available to us, such as InkOverlay and InkCollector. They each offer various advantages and disadvantages depending on the type of application you are developing. You can use the InkEdit control for applications when text recognition is required, such as address text boxes, and use the InkPicture control for scenarios when you don't need text recognition, such as when you need to capture signatures. The objects are equally easy to use and share many of the same properties, making your decision on which way to develop a very easy decision. In Chapter 14, Tablet PC Full Screen Utility, we look at some of the hardware of the Tablet PC, including its special buttons and screen rotation.

14 ▪ Tablet PC Full Screen Utility

I n the previous chapter, we built Tablet PC applications that introduced some of the common features we'll use throughout the book. Now, in this chapter, we are going to build our first useful application, a Tablet PC utility that resides in the taskbar and uses a context menu.

ON THE CD

The source code for the projects are located on the CD-ROM in the PROJECTS folder. You can either type them in as you go or you can copy the projects from the CD-ROM to your hard drive for editing.

FULL SCREEN UTILITY OVERVIEW

The application we build in this chapter allows a user to write anywhere on the screen (see Figure 14.1). Because the application encompasses the entire screen, the user needs a way to minimize and close the application. To do this, we'll test for the right mouse button being pressed. If the right button is pressed, the application will pause for a period of 2000 milliseconds. It will then send the entered text to whatever application currently has focus.

BUILDING THE APPLICATION

We begin this utility by creating a new Windows Forms application. As a utility, we're going to create this application so that it will run in the taskbar. The taskbar menu will have three options to choose from:

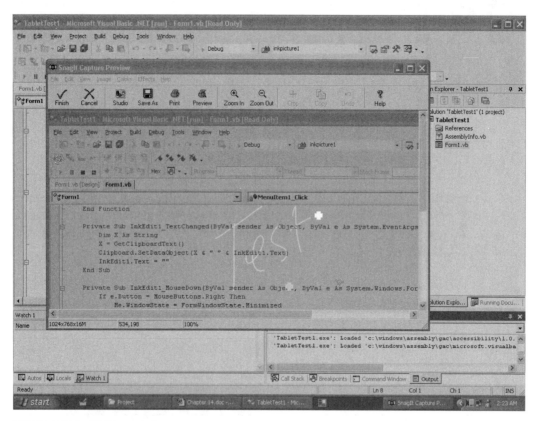

FIGURE 14.1 Writing on the screen.

Display: Fills the screen with our application

Paste: Copies the text to the clipboard as it's recognized

Exit: Closes the application

To create a taskbar menu in Visual Basic .NET, we can use a NotifyIcon control and a ContextMenu control. You can add these controls to our form. In the Properties window, you need to set the ContextMenu> property to Context Menu1. This displays the context menu automatically when the icon is right-clicked in the taskbar.

The next step is to add an InkEdit control to the form (see Figure 14.2). We'll use the InkEdit control because it automatically recognizes ink as we enter it. We could have used InkPicture and recognized the ink, or we could have used the API.

Any of these work, but the InkEdit control is the easiest for this application. Because we are going to position and resize the control programmatically to fill the entire screen, you don't need to worry about its position or size.

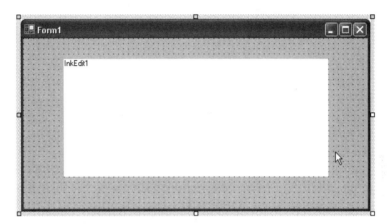

FIGURE 14.2   The control on the form.

Assigning Important Properties in Code

Let's write some code beginning with the Form_Load> event procedure. We'll set a few important properties for our application. For starters, our application is going to be transparent so that we can see which applications are running beneath it.

Here is the code for the Form_Load> event:

```
Private Sub Form1_Load(ByVal sender As System.Object, ByVal e As
System.EventArgs) Handles MyBase.Load

    With Me
        .TopMost = True
        .Opacity = 0.3
        .Text = ""
        .ControlBox = False
        .FormBorderStyle = FormBorderStyle.None
        .MaximizeBox = False
        .MinimizeBox = False
        .ShowInTaskbar = True
        .BackColor = System.Drawing.Color.Gray
    End With

    NotifyIcon1.Visible = True
```

```
        NotifyIcon1.Icon = Me.Icon

        NotifyIcon1.Text = "Full Screen Tablet Utility"

        Me.WindowState = FormWindowState.Minimized
        Me.Hide()

    End Sub
```

In the previous code, you can see that most of this procedure deals with setting properties for the form. This begins with setting the form to be located at the top of any running applications. This is obviously a very necessary feature for our application because we are planning to write on the entire screen and need to be above the other applications so we can capture the ink.

Next, we set the opacity form to 0.3. This allows us to "see through" the form so that other running applications are visible as we are inking. Next, we change the text property to an empty string and the ControlBox> property to False> so that a control box is not displayed in the caption of the form. We can then set the border style, maximize box, and minimize box properties so that when the application is maximized, it takes up the entire screen. The last line of the With> statement changes the background color to gray.

After setting the properties for the form, we turn our attention to the Notify-Icon. First, we set the Visible> property to True> so that we can see it, we assign the icon to be the same as the icon in the application (if we had wanted to, we could have picked any icon for this property so that it would have been seen in the taskbar), and lastly we assign the Text> property to "Full Screen Tablet Utility"> so that when a mouse points to it, the text is seen by the user.

In order to wrap up the Form_Load> procedure, we set the form to minimize so that it isn't displayed on startup and then we decide to hide the application so that it won't be seen at all.

We now create the ContextMenu entries. Back in the Form Designer, we can click on the ContextMenu icon. This displays the context menu in the form. Next, click on the area within the form that shows the ContextMenu text. When you click the menu, a new menu option can be seen called "Type Here." This is the first entry that will be seen by the end user. We can add the text "Display" to this. As you type this entry, new menus are available to its right and beneath it. We are going to add the following two items, in order, beneath the first one:

Paste

Exit

We now have three menu options (see Figure 14.3), but we need to create code for each of the menu events. You can create the click event for "Display" by double-clicking its entry in the menu. In the Code Editor, add the following code:

```
Private Sub MenuItem3_Click(ByVal sender As System.Object, ByVal e As
System.EventArgs) Handles MenuItem3.Click
    Me.WindowState = FormWindowState.Maximized
    Me.Show()
    Me.TopMost = True
    InkEdit1.Size = Me.Size
    InkEdit1.Text = ""
    InkEdit1.Left = 0
    InkEdit1.Top = 0
    InkEdit1.BackColor = System.Drawing.Color.Gray
    InkEdit1.DrawingAttributes.Color = System.Drawing.Color.White
End Sub
```

Display

Paste

Exit

FIGURE 14.3   The menu options.

The first line of code sets the window to maximize so that the entire screen is filled with the application's window. Next, we need to show the application so that it can be seen. We then make sure that it is set to the top position among running applications before we move on to set InkEdit properties, the first of which is the Size> property. Next, we set InkEdit to an empty string and position it at the top left of the form so that it now takes up the entire screen. We finish up the procedure by setting the background color of InkEdit to gray, and then set the color of the pen to white.

Next, create the event procedure for the Paste menu item and add the following code:

```
Private Sub MenuItem1_Click(ByVal sender As System.Object, ByVal e As
System.EventArgs) Handles MenuItem1.Click
    System.Threading.Thread.Sleep(2000)
    SendKeys.Send("^+V")
    Clipboard.SetDataObject("")
End Sub
```

The first line of code from the procedure delays the system 2 seconds. This gives us time to position our cursor before the text is pasted. To paste the text, we use the

SendKeys> method. There are several variations of SendKeys> and numerous "nondisplaying" commands we can use. We'll look at a few variations. First, we can use SendKeys>, which sends keystrokes to an active window of the foreground application. If you need further control, we can use SendKeys.SendWait>, which sends keystrokes and execution is suspended until the keystrokes have been processed.

To send normal alphanumeric characters, it's very easy. However, for our application, we need to send special keys that represent the Control key being pressed along with "V" to simulate the Paste command's shortcut key combination.

The following characters represent special keys:

+: Shift

^: Ctrl

%: Alt

To use the special keys, you surround the character with braces. For example, to specify the Alt key, you can use {%}. Preceding a standard string with the special characters described in the previous list allows you to send a keystroke combination beginning with Shift, Ctrl, or Alt. For example, to specify Ctrl followed by "J," use ^J. If you need to specify that the Shift, Ctrl, or Alt key is held down while another key is pressed, you should enclose the key or keys in parentheses and precede the parentheses with the special character code. For example, to specify the J key being pressed while holding down the Alt key, you would use %(J).

The following list describes how to specify "nondisplaying" characters:

Backspace: {BACKSPACE}, {BS}, or {BKSP}

Break: {BREAK}

Caps Lock: {CAPSLOCK}

Del or Delete: {DELETE} or {DEL}

Down arrow: {DOWN}

End: {END}

Enter: {ENTER}or ~

Esc: {ESC}

Help: {HELP}

Home: {HOME}

Ins or Insert: {INSERT} or {INS}

Left arrow: {LEFT}

Num Lock: {NUMLOCK}

Page Down: {PGDN}

Page Up: {PGUP}

Right arrow: {RIGHT}

Scroll Lock: {SCROLLLOCK}

Tab: {TAB}

Up arrow: {UP}

F1: {F1}

F2: {F2}

F3: {F3}

F4: {F4}

F5: {F5}

F6: {F6}

F7: {F7}

F8: {F8}

F9: {F9}

F10: {F10}

F11: {F11}

F12: {F12}

F13: {F13}

F14: {F14}

F15: {F15}

F16: {F16}

You can also use some formatting to simulate a key being pressed repeatedly. For example, {L 5} represents pressing the L key five times. With all of the information about SendKeys>, you can see that we are simulating Ctrl-V with the final line of code from the previous code listing.

We can now create the final menu event, which is the easiest of the three. If "Exit" is clicked, we need to simply end the application. Here is the code:

```
Private Sub MenuItem2_Click(ByVal sender As System.Object, ByVal e As
System.EventArgs) Handles MenuItem2.Click
    End
End Sub
```

CAPTURING INK

Our application is nearly finished, but we have yet to deal with ink collection. As ink is entered into InkEdit1 and then recognized, it will raise the TextChanged> event. We can take advantage of this to add information to the clipboard. As you already know, pasting from the clipboard is one feature of the application. We're also going to use the clipboard to store the text as we recognize it.

Let's create the TextChanged> event for InkEdit1 (choose InkEdit> as the Class Name and TextChanged> as the Method), and then add the following code:

```
Private Sub InkEdit1_TextChanged(ByVal sender As Object, ByVal e As
System.EventArgs) Handles InkEdit1.TextChanged
    Dim X As String
    X = GetClipboardText()
    Clipboard.SetDataObject(X & " " & InkEdit1.Text)
    InkEdit1.Text = ""
End Sub
```

The first line of code creates a variable that we can use to store the contents of the clipboard. If text content is on the clipboard, we need to store the text in the variable and then set the new clipboard data to be the original string followed by a space and then the newly recognized text. The following function is the GetClipboardText> function that we called in the previous procedure:

```
Public Function GetClipboardText() As String
    Dim objClipboard As IDataObject = Clipboard.GetDataObject()
    With objClipboard
        If .GetDataPresent(DataFormats.Text) Then Return _
            .GetData(DataFormats.Text)
    End With
End Function
```

You can see that the code for the function is very straightforward. We begin by retrieving data and checking if the information is a text string. If so, we return the string; otherwise, we make sure nothing is returned.

The final part of the application that we need to handle is a way to minimize the application once it has taken over the entire screen. Because the application limits the ability of the pen or mouse to click a button from the taskbar to close it, we need another way to minimize the form. One way is to add a button to the form, but we want to leave the entire screen available for ink. The remaining way we could minimize the application is to test the use of the right mouse button. If it is clicked, we can minimize the form and then hide the application. We can also add the ability to automatically paste the contents stored onto the clipboard. We've already created code to handle this, so we can simply call MenuItem1 (your name may be different depending on the order you used to create the menu options).

Here is the code for the procedure:

```
Private Sub InkEdit1_MouseDown(ByVal sender As Object, ByVal e As
System.Windows.Forms.MouseEventArgs) Handles InkEdit1.MouseDown
    If e.Button = MouseButtons.Right Then
        Me.WindowState = FormWindowState.Minimized
        Me.Hide()
        MenuItem1.PerformClick()
    End If
End Sub
```

You can now test the application and then see how it works (see Figure 14.4).

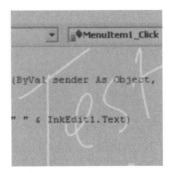

FIGURE 14.4 The application is being tested.

SUMMARY

In this chapter, we created a utility that allows a user to write anywhere on the screen. The ink is converted into text and is placed on the clipboard so that it can be pasted into any application. The utility could use a few extra features, such as the ability to enable or disable the automatic pasting, but it is fully functional at this time. We create several additional utilities later in the book. The utilities are similar enough to this one that you can come back to add the missing features.

15 Tablet PC Screen Rotation and Special Buttons

In the previous chapter, we built an application that allowed us to write any-where on the screen and then store the recognized text on the clipboard. The text was then pasted automatically or manually into another running applica-tion. In this chapter, we turn away from ink collection for a moment and look at another interesting aspect of the Tablet PC—its special buttons and rotation of the screen (see Figure 15.1).

ON THE CD

The source code for the projects are located on the CD-ROM in the PROJECTS folder. You can either type them in as you go or you can copy the projects from the CD-ROM to your hard drive for editing.

EXTENDING VB .NET WITH C++

VB is the most popular programming language in the world for a variety of reasons. There are many great resources for learning the language and it has become in-creasingly popular with each release. However, there are times when you are tack-ling a problem when C++ still makes the most sense. We happen to have one of those situations as we look at programmatically rotating the screen. Fortunately, it is a nearly painless process to create a DLL in C++ that we can call from VB or C#. In case you don't have any experience with C++, the DLL is only a few lines of code.

ON THE CD

If you don't want to build the DLL, you can use the one included on the CD-ROM in the Chapter 15 folder.

Begin by opening VS .NET. Choose Visual C++ Project from the list of project types and then choose Win32 Project from the available list. Similarly to VB, type the name of the DLL in the Name field. This name is important because it will be used by your VB .NET program when you call the DLL from it. Therefore, choose something like "screenrotatedll." We'll assume this is the name the rest of the way,

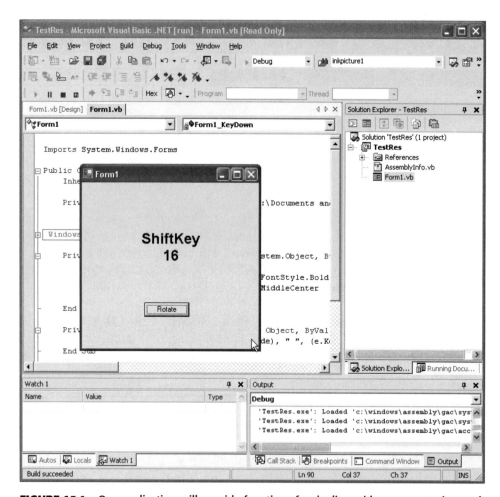

FIGURE 15.1 Our application will provide functions for dealing with screen rotation and special buttons.

but you can replace this name with anything you want. Also, remember the location of the project because we'll use the path in the VB project as well.

After you create a Win32 Project, you can choose the Application Settings tab (see Figure 15.2), and then select DLL as the type of application. Double-click on the Screenrotatedll.cpp file to open it. This is the main code for our project, and although we could create new cpp and h documents, it's easier for this simple project to use the existing cpp file. When you open it, you'll see there is a single function defined for our DLL. You can think of this function as being equivalent to the Form_Load event in VB .NET.

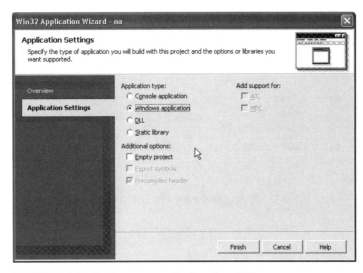

FIGURE 15.2   Setting our application to a DLL.

The code will look similar to the following, but will be slightly different depending on what you named your project:

```
// screenrotatedll.cpp : Defines the entry point for the DLL
application.
//

#include "stdafx.h"
BOOL APIENTRY DllMain( HANDLE hModule,
                       DWORD  ul_reason_for_call,
                       LPVOID lpReserved
                     )
{
    return TRUE;
}
```

Next, we need to add the WinGDI header file to our project. Choose Add Existing Item from the Project menu and then navigate to the following folder: C:\Program Files\Microsoft SDK\include.

Within this folder, you should find the Win32GDI.h file. Choose this file and then click Open. You will see it is available in Solution Explorer, located in the Header Files folder. Now, we need to include it in the project. Add the following code:

```
#include "WinGDI.h"
```

This `include` should be placed immediately beneath the existing `include` statement. This simply includes the WinGDI header file so that we have access to it in our project. At the end of the current code, you will see the following:

```
{
Return True
}
```

Place your cursor after the final bracket and press Enter to place it beneath all existing code. This is where we will place all of our new code, beginning with our only method, `rotscreen`.

Here is the complete code for the method:

```
int _stdcall rotscreen(int value){

DEVMODE dm;
DWORD dwTemp;
EnumDisplaySettings(NULL, ENUM_CURRENT_SETTINGS, &dm);
dwTemp = dm.dmPelsHeight;
dm.dmPelsHeight = dm.dmPelsWidth;
dm.dmPelsWidth = dwTemp;

if (value == 90) {
    dm.dmDisplayOrientation = DMDO_90;
}

if (value == 180) {
    dm.dmDisplayOrientation = DMDO_180;
}

if (value == 270) {
    dm.dmDisplayOrientation = DMDO_270;
}

dm.dmFields = DM_DISPLAYORIENTATION|DM_PELSHEIGHT|DM_PELSWIDTH;
ChangeDisplaySettings(&dm, 0);
return value;
}
```

The previous code uses the `ChangeDisplaySettings` API. We are not going to really get involved with explaining this too much. If you want to learn more about this, there is information available in the Platform SDK, which you can find at *http://www.microsoft.com/msdownload/platformsdk/sdkupdate/.*

If you have ever worked with some basic C++, you may be wondering about the _stdcall keyword. This is added to make sure that the function is used in the same way as VB for passing variables. The VB equivalent for what we have is as follows:

```
Function rotscreen() As Integer
        ...code
End Function
```

Before we can use the DLL in VB, we must make sure that the function we created is viewable by outside applications. To do this, we use a special type of file called a Definition File (DEF), which is actually very simple. It is nothing more than a list of function names to export so applications know what they're looking for as it queries the DLL for methods. To add a new file to the project, select Project | Add New Item | DEF file (.def). In the Name field, type the name of your project (i.e., screenrotatedll), and then click the Open button.

We now have the DEF file available in our project and it is automatically opened in the IDE. You should see a single line at the top of the code:

```
LIBRARY     screenrotatedll
```

This reports the screenrotatedll name to other applications, but we still need to add the export. For our application, this is a single method called rotscreen. We can use the keyword EXPORTS:

```
LIBRARY     screenrotatedll
EXPORTS
    rotscreen
```

Now, we need to compile the project. Choose Build screenrotatedll from the Build menu. You will see the build log being generated as the files are compiled and linked, and when finished, you should see something like:

```
------ Rebuild All started: Project: screenrotatedll, Configuration:
Debug Win32 ------

Deleting intermediate files and output files for project
'screenrotatedll', configuration 'Debug|Win32'.
Compiling...
stdafx.cpp
Compiling...
screenrotatedll.cpp
Linking...
   Creating library Debug/screenrotatedll.lib and object
Debug/screenrotatedll.exp
```

```
Build log was saved at "file://c:\Documents and Settings\Clayton\My
Documents\TabletPC\Chapter 15\Project\screenrotatedll\Debug\
BuildLog.htm"
screenrotatedll - 0 error(s), 0 warning(s)


--------------------- Done ---------------------

        Build All: 1 succeeded, 0 failed, 0 skipped
```

If you receive any error messages or warnings, you need to go back through the code to make sure you didn't miss something, such as a ";". With this small amount of code, it should be relatively easy.

USING THE DLL IN VB .NET

If you have ever used the API in VB .NET, you are already familiar with our next step, which is to use the rotscreen function in VB. You can close the C++ project and create a new VB .NET Windows Forms application. At the top of the code (beneath the Inherits System.Windows.Forms.Form line), we need to add the following code:

```
Private Declare Function rotscreen Lib "C:\Documents and Settings\
Clayton\My Documents\TabletPC\Chapter 15\Project\screenrotatedll\
Debug\screenrotatedll.dll" (ByVal value As Integer) As Integer
```

You can see the long path location, and you can copy your DLL to the root drive or even the System folder. Wherever you put it, you need to change the path to represent the actual location. You might wonder what the rest of this means. The following list is a breakdown of the syntax.

"Private": The function is only available here.

"Declare": This means that this is only a method header and not the entire method.

"Function": It's a function.

"Lib": This is the library name given to VB.

"screenrotatedll.dll": This is the name of the DLL file that contains the method. This is an extremely long path in our example, but you can change it to something manageable.

"(": We're going to declare some parameters that need to be passed to the method.

"ByVal": We're passing the value of the variable passed to the method and not the address of its value in memory.

"As Integer": This means we're setting the type of value being passed to an integer.

")": This is the end of the parameters.

"As Integer": This returns an integer.

That's really all we need to add to the program to provide access to the `rotscreen` function. Now, let's add the following two controls to the form, as shown in Table 14.1.

TABLE 14.1 Adding Label and Button controls to the form

Type	Name	Text
Label	lblKeys	
Button	btnRotate	Rotate

The next step is to set a few properties to the `Form_Load` event:

```
lblKeys.Text = ""
lblKeys.Font = New Font("Arial", 18, FontStyle.Bold)
lblKeys.TextAlign = ContentAlignment.MiddleCenter
Me.KeyPreview = True
```

We begin by setting the `Text` property of `lblKeys` to an empty string, and then we set its font and text alignment properties so that it is more easily seen on the form. The label will display the key as it is being pressed. The keys can be either the standard keyboard keys, or the special Tablet PC keys that are included in various locations, depending on the model of Tablet PC. These keys include things such as Tab and Escape. The final property sets the `KeyPreview` property of Form1 to be `True`. This lets the form handle the `keydown` event. Without this, the form would ignore the key presses.

Because we have this in mind, let's create the `Form1_KeyDown` event and add the following code:

```
lblKeys.Text = String.Concat((e.KeyCode), " ", (e.KeyValue))
```

This line of code displays the key being pressed and the value of the key. It uses the label we already have on the form.

The final thing we need to do with our application is to rotate the screen. We can place this code in the btnRotate_Click event. After you have created the procedure, add the following code:

```
Dim ok As Integer
ok = rotscreen(90)
```

You may have noticed that VB automatically provides us help for our rotscreen function as you type it, just like it does for any DLL. As you can see in Figure 15.3, when you type the rotscreen function name, VB informs us that it is looking for an integer to be passed and that it will return an integer. For our application, we need to pass values such as 90, 180, or 270, so that the screen can be rotated. Any other type of value being passed causes the function to return the value being passed and does nothing to the screen.

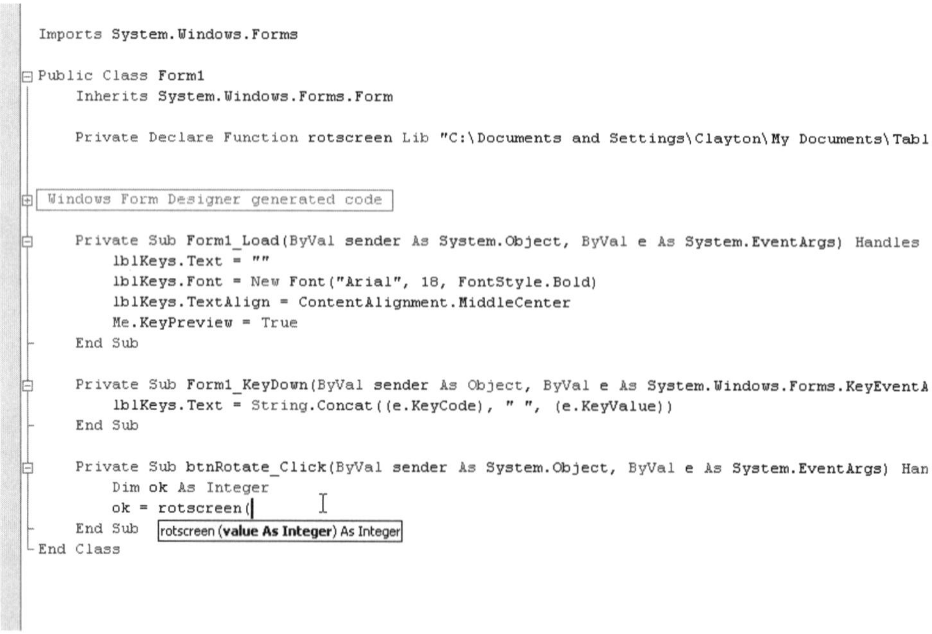

```
Imports System.Windows.Forms

Public Class Form1
    Inherits System.Windows.Forms.Form

    Private Declare Function rotscreen Lib "C:\Documents and Settings\Clayton\My Documents\Tabl

    Windows Form Designer generated code

    Private Sub Form1_Load(ByVal sender As System.Object, ByVal e As System.EventArgs) Handles
        lblKeys.Text = ""
        lblKeys.Font = New Font("Arial", 18, FontStyle.Bold)
        lblKeys.TextAlign = ContentAlignment.MiddleCenter
        Me.KeyPreview = True
    End Sub

    Private Sub Form1_KeyDown(ByVal sender As Object, ByVal e As System.Windows.Forms.KeyEventA
        lblKeys.Text = String.Concat((e.KeyCode), " ", (e.KeyValue))
    End Sub

    Private Sub btnRotate_Click(ByVal sender As System.Object, ByVal e As System.EventArgs) Han
        Dim ok As Integer
        ok = rotscreen(
    End Sub        rotscreen (value As Integer) As Integer
End Class
```

FIGURE 15.3   Help in the IDE is provided automatically.

You can now save this application and test the various features. Figures 15.4 and 15.5 display some of its capabilities.

FIGURE 15.4   The form has
detected a key being pressed.

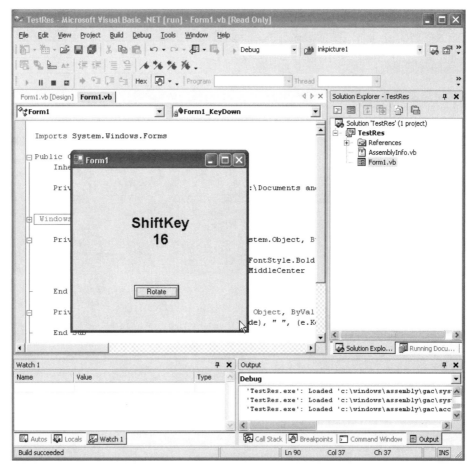

FIGURE 15.5   The screen is rotated by 90 degrees.

SUMMARY

In this chapter, we looked at how we can easily extend VB .NET with C++ DLLs and have seen that collecting ink is not the only important aspect of a Tablet PC application. The screen rotation DLL that we created can be very useful from a development standpoint, as one of the interesting features offered by the Tablet PC is screen rotation.

16 Creating an MP3 Player

The popularity of MP3 music is unquestioned. There have been several lawsuits directed by the recording industry at companies that aid users in sharing the files, which many times are protected under copyright laws. The media attention to the format is incredible, and with all the publicity, the user base of programs like Kazaa, Morpheus, or any of the similar offerings that allow individuals to freely exchange MP3 files, continues to grow nearly exponentially. This is not an endorsement of exchanging copyrighted materials, but rather an acknowledgment that regardless of the outcome of future lawsuits, the format itself is so popular that it will undoubtedly remain in one capacity or another. With this in mind, developing an MP3 player is a very good project for learning Visual Basic and Tablet PC programming concepts.

ON THE CD

The source code for the projects are located on the CD-ROM in the PROJECTS folder. You can either type them in as you go or you can copy the projects from the CD-ROM to your hard drive for editing.

PROJECT OVERVIEW

In this chapter, you'll develop an MP3 player. This project could be based around a number of commercial ActiveX controls, but the goal of this book is to allow you to complete most of the projects without spending any additional money.

Instead of an ActiveX control, you could also remotely control another application from a Visual Basic program. For instance, it is relatively easy to write a program that remotely controls an existing MP3 player. Again, this option requires you to purchase additional software, which you shouldn't have to do. Another problem with this approach occurs if you decide to distribute your application to other users. Controlling another application requires the user of your MP3 player to purchase another MP3 player. Instead of an ActiveX control or another application, the solution we use is the Winmm.dll file and Multimedia Control Interface (MCI) commands.

THE PROJECT

To begin our project, we will construct the basic GUI for the MP3 player. Start the Visual Basic IDE, and from the New Projects window, make sure that Visual Basic is selected and then select Windows Application. You can change the name of the project to something you'd like (it is named Chapter 16 on the CD-ROM and throughout the chapter).

You need to place several controls on the form, one of which is the Open File Dialog. You can place this control anywhere you want as it's not visible at runtime. When you drag it onto the form, it will be displayed beneath the form (see Figure 16.1).

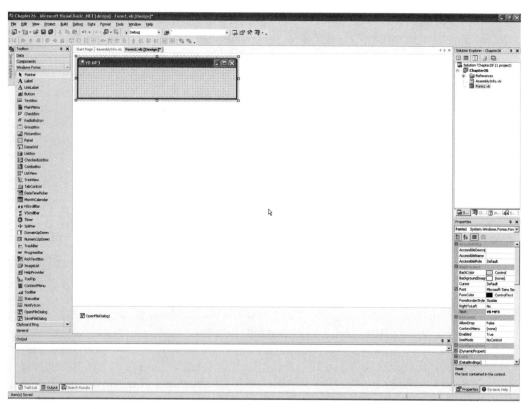

FIGURE 16.1 The File Open Dialog visible beneath the form and with the Text Property changed.

The next step is to alter the Text property of the form to read "VB MP3" (without quotations). You will find this property in the Properties window. Next, place a series of command buttons for opening, stopping, and so on. The buttons can be arranged along the top side of the form going from left to right and can be named as follows: cmdOpen, cmdPlay, cmdStop, cmdPause, and cmdClose. Their captions should also be altered to reflect their intended usage; that is, the cmdOpen button should have its caption property renamed to Open. When it's finished, the form should look similar to Figure 16.2.

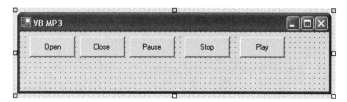

FIGURE 16.2 The interface is beginning to take shape with command buttons in place.

The final components that need to be added are a Label control, which will be used to display the filename of the song, and another Label control, which is added to identify the use of the control to the end user. You can name the control lblCaption and change the Text property to an empty string (just erase anything inside the field in the Properties window). The last Label control can be positioned to its left and given a Text Property of "Filename :".

If a specific name isn't given, you can use the standard VB assigned names.

Your final interface should appear similar to Figure 16.3.

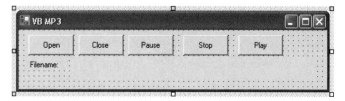

FIGURE 16.3 The final GUI.

SETTING THINGS UP

This application uses several variables to store information. If you double-click the form, it displays the Code Editor and your cursor will be visible in the `Form1_Load` event. You can move the cursor directly above this event and place the following variable declarations directly beneath "`Inherits System. Inherits. System.Windows.Forms.Form`":

```
Private Declare Function mciSendString Lib "winmm.dll" _
Alias "mciSendStringA" (ByVal lpstrCommand As String, _
ByVal lpstrReturnString As String, ByVal uReturnLength _
As Long, ByVal hwndback As Long) As Long

    Dim strFileName As String
    Dim blnPlaying As Boolean
    Dim Temp As Integer
    Dim command As String
    Dim s As String
    Dim strFileNameTemp As String
```

You may have noticed the underline character "_" located in the `Private Declare` *statement. This simply allows you to use multiple lines to display a line that should actually be located in a single line.*

Although there are several variables, the first line, which begins with `Private Declare . . .`, is the most interesting. It's a Windows Application Programming Interface (API) call. For now, you don't need to concern yourself too much with API calls as they are discussed in detail in Chapter 17, Form Effects, but you do need to understand that the `mciSendString` function is now available to the entire form. The `mciSendString` function sends a command to a Multimedia Control Interface (MCI) device. The command strings used with this function can perform almost any task necessary for using a multimedia device installed on the computer and provides a relatively easy way to perform multimedia output operations.

The `Form1_Load` event is called when the form is first displayed. We'll use this event to set up the lblCaption to read " – No Media". Enter the following code for the `Form_Load` event, which sets lblCaption to a value to inform the user that nothing is currently opened:

```
Private Sub Form1_Load(ByVal sender As System.Object, _
    ByVal e As System.EventArgs) Handles MyBase.Load

    lblCaption.Text = " -- No Media"

End Sub
```

PLAYING THE MP3 FILES

You have created a command button called cmdOpen that will be used by the program to open the MP3 file. We simply use the cmdOpen_Click event, which occurs when the button is clicked. Before we play the file, we need to check to see if a file is already playing, and if there is one playing, you should exit the procedure without opening a file. However, you should let the user know that they are trying to open the player when a file is already open. If a file is not playing, you should continue on with opening a file, which begins with initializing the OpenFileDialog control so that it displays only MP3 files. Lastly, you should use the mciSendString function to open the file.

The following code does all of this:

```
Private Sub cmdOpen_Click(ByVal sender As System.Object, _
ByVal e As System.EventArgs) Handles cmdOpen.Click


        On Error GoTo ErrorHandler
        If blnPlaying Then
            MsgBox("Player is Busy!", vbExclamation)
            Exit Sub
        End If
        OpenFileDialog1.Filter = "MP3 Files|*.MP3"
        OpenFileDialog1.ShowDialog()

        If OpenFileDialog1.FileName = "" Or _
OpenFileDialog1.FileName = strFileName Then

        Else

            strFileName = OpenFileDialog1.FileName
            strFileNameTemp = OpenFileDialog1.FileName
            FileCopy(strFileName, "C:\mciplay")
            strFileName = "C:\mciplay"

            mciSendString("open " & strFileName & " type MPEGVideo", _
    0, 0, 0)
            lblCaption.Text = strFileNameTemp
        End If
ErrorHandler:
    End Sub
```

There is an interesting problem with using the mciSendString method for playing MP3 audio files. The filename cannot be more than 8 characters followed by 26 characters as an extension. It's obviously not practical to limit your application to use only files named in this format, so we'll solve this by copying the file to the root directory of your hard drive and giving it a name of MCIPlay. We'll then use this filename to actually play the file. The `mciSendString` command is very strict about its syntax, uses quotation marks, and includes MPEGVideo as its type. Although it says Video, it is the correct type for MP3 audio files as well.

THE REST OF THE FUNCTIONS

After you have opened the file, you can use the command buttons you created for playing, stopping, pausing, and so on. The programming that is used in all of the procedures is nearly identical. First, you need to see if a file is already playing by checking the `blnPlaying` variable, which is of type Boolean. A Boolean variable can display only two values "True" or "False"(could also be thought of as 0 or 1, Off or On). If a file is playing, you can send an MCI code to do the task. Lastly, you can change the lblCaption to reflect the command.

The following lists all of the procedures:

```
Private Sub cmdClose_Click(ByVal sender As System.Object, _
 ByVal e As System.EventArgs) Handles cmdClose.Click
        If blnPlaying Then
            mciSendString("close " & strFileName, 0, 0, 0)
        End If
        blnPlaying = False
        lblCaption.Text = " --No Media"
End Sub

Private Sub cmdPause_Click(ByVal sender As System.Object, _
 ByVal e As System.EventArgs) Handles cmdPause.Click
        If blnPlaying Then
            mciSendString("pause " & strFileName, 0, 0, 0)
            blnPlaying = False
            lblCaption.Text = OpenFileDialog1.Filename & " --Paused"
        End If
End Sub

Private Sub cmdPlay_Click(ByVal sender As System.Object, _
 ByVal e As System.EventArgs) Handles cmdPlay.Click
        If strFileName <> "" Then
            mciSendString("play " & strFileName, 0, 0, 0)
            blnPlaying = True
            lblCaption.Text = OpenFileDialog1.Filename & " -- Playing"
```

```
            End If
    End Sub

    Private Sub cmdStop_Click(ByVal sender As System.Object, _
     ByVal e As System.EventArgs) Handles cmdStop.Click
            If blnPlaying Then
                mciSendString("stop " & strFileName, 0, 0, 0)
                blnPlaying = False
                lblCaption.Text = OpenFileDialog1.Filename & " -- Stopped"
            End If
    End Sub
```

The only procedure that differs slightly is the cmdPlay_Click procedure, which does not check the status of a playing file. Rather, it only needs to determine if the strFileName variable contains information. If it does, it knows that it must have a file open. If not, it exits the procedure.

Figure 16.4 represents what the finished MP3 player should look like when playing a file. If you are interested, you can add common functions such as play lists or captions that display time-related information. This project is used in Chapter 17, Form Effects, to add some interesting changes to the vanilla-looking GUI.

FIGURE 16.4 Playing an MP3 file.

ADDING TABLET FUNCTIONALITY

At this point, the application doesn't really offer any features that would make it specifically useful for a Tablet PC. In this section, we add some functions that allow us to open, close, and pause the player by using only the pen. Obviously, because the stylus acts as a mouse, you could click the buttons to execute their respective tasks. However, we are going to write commands directly to the application to perform the tasks. You'll see that this isn't really much of a time-saver for a pen user, and we'll look at a shorthanded version of pen input, known as gestures in Chapter 18, Using Gestures to Control Tablet Media Player.

There are a number of ways in which we could proceed for a method to allow pen input, but perhaps the easiest way is to hide the command buttons we used to create the application by placing an InkEdit control over them as seen in Figure 16.5.

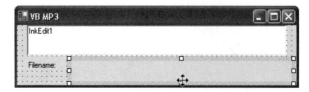

FIGURE 16.5 The InkEdit control roughly hides the command buttons.

Now that the buttons are not visible, we need to have a way to control the application functions via text entry. The first thing we need to do is change the Text property of the InkEdit control to an empty string using the Form_Load event:

```
InkEdit1.Text = ""
```

The remaining changes required for this application are very simple. We'll take advantage of the PerformClick method available for the existing buttons. Instead of requiring us to rewrite the existing code, we can simulate the buttons being pressed, which runs the code we have already written.

We'll use a Case statement to browse through the various options for checking the commands that are being entered into InkEdit. We can use the TextChanged event for InkEdit in which we take text from InkEdit, convert it to text, and then check this for the various commands (i.e., Play, Pause, and so on). We'll simulate the correct button press depending on the text entered into the InkEdit control. After we go through the various options, we need to set the InkEdit Text property to an empty string.

Here is the code:

```
Private Sub InkEdit1_TextChanged(ByVal sender As Object, ByVal e As
System.EventArgs) Handles InkEdit1.TextChanged
    Select Case InkEdit1.Text.ToUpper
        Case "OPEN " ' needs space after this or could remove
            cmdOpen.PerformClick()
        Case "CLOSE "
            cmdClose.PerformClick()
        Case "PLAY "
            cmdPlay.PerformClick()
        Case "STOP "
            cmdStop.PerformClick()
```

```
        Case "PAUSE "
            cmdPause.PerformClick()
        Case "EXIT"
            End
        Case Else

    End Select
    InkEdit1.Text = ""
End Sub
```

Our first step is to add a reference to the Tablet PC API (and Tablet PC API 1.5 if you want). Next, add the following declarations:

```
Public myInk() As Byte
Public myInkString As String
Dim WithEvents theInkCollector As InkCollector
```

SUMMARY

In this chapter, we built a fully functional MP3 player and have used an InkEdit control to handle the functions it provides. You sent commands to the Winmm.dll file by using mciSendStrings, which played the MP3 files.

17 Form Effects

By itself, Visual Basic provides the necessary basic ingredients for Graphical User Interface (GUI) design. Unfortunately, although you can change the color and shapes of items such as forms and controls, you're severely limited to the variations you can design. In this chapter, you'll create an application that doesn't really have a great deal of functionality, but instead focuses on designing a unique interface using the new GDI+ features of VB .NET that we initially looked at in Chapter 6, Object-Oriented Programming with VB .NET. As you already know, a Tablet PC runs what is essentially Windows XP. With that in mind, it's important to have an understanding of the new GDI+ if you want to take advantage of the many features a Tablet PC has to offer. In fact, in Chapter 26, Pong Game, we'll look at game development with a Tablet PC, a market that thus far is pretty limited.

ON THE CD

The source code for the projects are located on the CD-ROM in the PROJECTS folder. You can either type them in as you go or you can copy the projects from the CD-ROM to your hard drive for editing.

THE VB .NET WAY

Before looking at the VB .NET approach, it's worth rehashing (or mentioning for the first time if you skipped over the section) the VB6 method. To begin, we had to use several relatively obscure API calls as VB6 is fairly limited in built-in graphics features. On the other hand, .NET offers the GDI+ namespace that contains basically everything we need for this project.

We begin this project by opening the project from the previous chapter. We will use the same code for all of the playing and so on, but we are going to change the way the form looks. We begin by adding the following line:

```
Imports System.Drawing.Drawing2D
```

We'll also add a new variable of type `Point` beneath the variables from the previous chapter (the others are listed for your convenience):

```
Dim strFileName As String
Dim strFileNameTemp As String
Dim blnPlaying As Boolean
Dim Temp As Integer
Dim command As String
Private mouse_offset As Point
```

In the `Form_Load` event, we need to add a little new code. First, we'll change the opacity of the form to a value of .75 so that you can see through it. This is one of the more powerful new features offered to us. By changing the opacity property, we can produce in a single line what would be difficult in earlier versions of VB.

Here is the line of code:

```
Me.Opacity = 0.75
```

We'll now create regions and points for drawing the shape of the form. We'll set a series of points to draw the shape, change the `BackColor` property of cmdOpen and cmdClose to navy blue, and we'll also set `cmdOpen.Text` equal to an empty string.

Here is the entire `Form_Load` procedure:

```
Private Sub Form1_Load(ByVal sender As System.Object, ByVal e As
System.EventArgs) Handles MyBase.Load
    lblCaption.Text = " — No Media"
    Me.Opacity = 0.75 ' New Code

    Dim windowRegion As Region
    Dim regionPoints(5) As Point
    Dim regionTypes(5) As Byte
    regionPoints(0) = New Point(0, 0)
    regionPoints(1) = New Point(Me.Width, 0)
    regionPoints(2) = New Point(Me.Width — 50, Me.Height)
    regionPoints(3) = New Point(50, Me.Height)
    regionPoints(4) = New Point(0, 0)

    Dim Cnt As Long
    For Cnt = 0 To 5
        regionTypes(Cnt) = PathPointType.Line
    Next Cnt
```

```
    Dim regionPath As New GraphicsPath(regionPoints, regionTypes)
    Me.Region = New Region(regionPath)

    cmdOpen.Text = ""  ' Remove existing text
    cmdOpen.BackColor = Color.Navy
    cmdClose.BackColor = Color.Navy

End Sub
```

The next step is to create the Form1_Paint event. You already know how to create the event using the drop-down lists. We're going to use the Paint event to draw
a gradient background from red to yellow in the form.
Here is the code:

```
Private Sub Form1_Paint(ByVal sender As Object, ByVal e As
System.Windows.Forms.PaintEventArgs) Handles MyBase.Paint
    Dim r As Rectangle = New Rectangle(0, 0, Me.Width, Me.Height)
    Dim g As Graphics = e.Graphics
    Dim lb As LinearGradientBrush = New LinearGradientBrush(r,
Color.Red, Color.Yellow,
LinearGradientMode.BackwardDiagonal)
    g.FillRectangle(lb, r)
End Sub
```

We now need to remove the border from the form by changing the FormBor-
derStyle property of the form to None in the Properties window. This removes the
border and the Close button that is at the upper right of most Windows applications and looks like an "x". We will add a new button to the form, give it a Text
property of "x" and then place it on the form. You'll see an example of this in Figure 17.1.

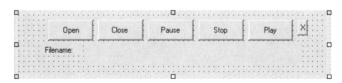

FIGURE 17.1 A new button added to the form.

Now that we have the button, we need to add the Click event for it and instruct
the application to close when it is clicked:

```
Private Sub cmdExit_Click(ByVal sender As System.Object, ByVal e As
System.EventArgs) Handles cmdExit.Click
    Me.Close()
End Sub
```

Because we removed the border, we will not have a built-in way to move the form by clicking and dragging it. Instead, we'll use the MouseDown and MouseMove events of the form to do this for us:

```
Private Sub Form1_MouseDown(ByVal sender As Object, ByVal e As
System.Windows.Forms.MouseEventArgs) Handles MyBase.MouseDown
    mouse_offset = New Point(-e.X, -e.Y)
End Sub

Private Sub Form1_MouseMove(ByVal sender As Object, ByVal e As
System.Windows.Forms.MouseEventArgs) Handles MyBase.MouseMove
    If e.Button = MouseButtons.Left Then
        Dim mousePos As Point = Control.MousePosition
        mousePos.Offset(mouse_offset.X, mouse_offset.Y)
        Location = mousePos
    End If
End Sub
```

We now have a form that looks much different than a standard VB form, but we're not finished. We're going to utilize the cmdOpen_Paint event and the cmd-Close_Paint event to change the way those two buttons work. If we were really interested in completing the application, we would follow the same procedure for the remaining buttons, but for this example, the two buttons work just fine.

Begin by creating both events. Next, we need to instantiate a new instance of the GraphicsPath class in the cmdOpen_Paint event:

```
Dim myGraphicsPath As New System.Drawing.Drawing2D.GraphicsPath()
```

Next, we need to specify a string that we'll draw. In this case, use "Open" and specify the font family to be "Arial" and the font style to be "Bold":

```
Dim stringText As String = "Open"
Dim family As FontFamily = New FontFamily("Arial")
Dim fontStyle As FontStyle = fontStyle.Bold
```

Our next step is to specify the size of an imaginary square, which will be used to house the string we are going to draw and a point at which the text will start. We'll use a size of 20:

```
Dim emSize As Integer = 20
Dim origin As PointF = New PointF(0, 0)
```

The last steps are to create a `StringFormat` object, which specifies the text formatting information, such as line spacing and alignment. We also need to use the `AddString` method to create the string, and lastly, we need to set the control's `Region` property to the instance of the `GraphicsPath` class we created earlier:

```
Dim format As StringFormat = StringFormat.GenericDefault

        myGraphicsPath.AddString(stringText, family, fontStyle, emSize,
origin, format)

        cmdOpen.Region = New Region(myGraphicsPath)
```

FINAL CODE LISTING

Here is the complete listing for both the `cmdOpen_Paint` event and `cmdClose_Paint` event:

```
        Private Sub cmdOpen_Paint(ByVal sender As Object, ByVal e As
System.Windows.Forms.PaintEventArgs) Handles cmdOpen.Paint
        Dim myGraphicsPath As New
System.Drawing.Drawing2D.GraphicsPath()

        Dim stringText As String = "Open"
        Dim family As FontFamily = New FontFamily("Arial")
        Dim fontStyle As FontStyle = fontStyle.Bold
        Dim emSize As Integer = 20

        Dim origin As PointF = New PointF(0, 0)

        Dim format As StringFormat = StringFormat.GenericDefault

        myGraphicsPath.AddString(stringText, family, fontStyle, emSize,
origin, format)

        cmdOpen.Region = New Region(myGraphicsPath)

    End Sub


        Private Sub cmdClose_Paint(ByVal sender As Object, ByVal e As
System.Windows.Forms.PaintEventArgs) Handles cmdClose.Paint
```

```
        Dim myGraphicsPath As New
System.Drawing.Drawing2D.GraphicsPath()

        Dim stringText As String = "Close"

        Dim family As FontFamily = New FontFamily("Arial")

        Dim fontStyle As FontStyle = fontStyle.Bold

        Dim emSize As Integer = 20

        Dim origin As PointF = New PointF(0, 0)

        Dim format As StringFormat = StringFormat.GenericDefault

        myGraphicsPath.AddString(stringText, family, fontStyle, emSize,
origin, format)

        cmdClose.Region = New Region(myGraphicsPath)
    End Sub
End Class
```

TESTING THE APPLICATION

We don't really need to test the application because the basics of file playback were tested in the previous chapter. We only need to run it to see if it is now drawn appropriately. We can also test the ability of the program to be moved with the mouse to make sure you can drag it around the screen. You can see an example of the form in Figure 17.2.

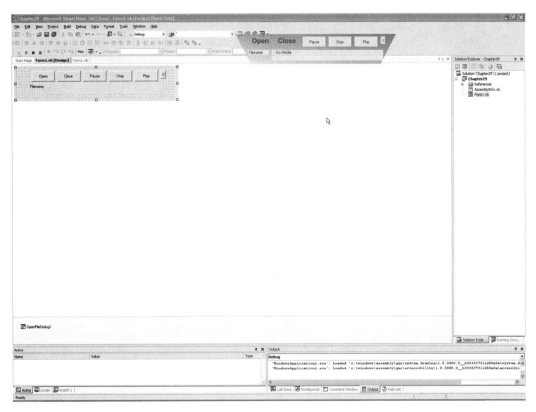

FIGURE 17.2   The final application.

SUMMARY

In this chapter, we built a custom interface for our MP3 player using GDI+. We use GDI+ functions in several chapters in the book, so you can refer back to this information if needed. In Chapter 18, Using Gestures to Control Tablet Media Player, we look at using gestures in a similar type of application.

18 Using Gestures to Control Tablet Media Player

If you recall from Chapter 16, Creating an MP3 Player, we created an MP3 player that allows the user to write out a command that is recognized and a certain behavior is executed, such as playing or stopping a file. Although it obviously works, it is definitely not an easy solution from a user standpoint, as the need to write the word "play" to play a file is very tedious. In this chapter, we look at a much better way to control these types of applications by using gestures to control a media player (see Figure 18.1).

ON THE CD

The source code for the projects are located on the CD-ROM in the PROJECTS folder. You can either type them in as you go or you can copy the projects from the CD-ROM to your hard drive for editing.

TABLET GESTURES

If you have ever used a PalmPilot, you have seen the way a device can take advantage of a type of shorthand called graffiti. The Palm works by allowing the user to write a simple version of complex characters—making entering text an easier objective. The Tablet PC provides a much deeper implementation of a shorthand system known as gestures.

The gestures are available as part of the standard Tablet PC Platform SDK. When you develop an application, the gestures are not enabled by default. Rather, if you want to enable one of the built-in gestures, you must instruct the application that you are doing so. Most gestures have a "hot point," which is a distinguishing point in the geometry of the gesture. The hot point can be used to determine where the gesture was performed. The gestures APIs provide the HotPoint property of the Gesture event, making it possible to determine the hot point for a given gesture. Not all gestures have a specific hot point; for these gestures, the starting point is reported as the hot point. Using both Figure 18.1 and Table 18.1, you can see an example of the gestures and the location of the hot point.

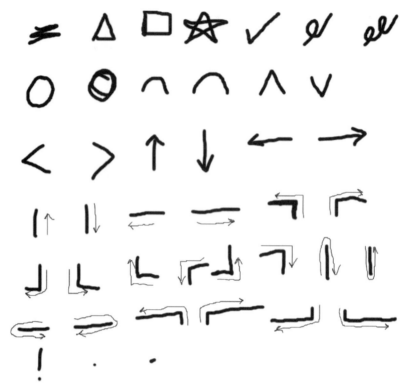

FIGURE 18.1 Built-in gestures from Table 18.1, viewed from left to right.

TABLE 18.1 Gestures

Name	Hot Point	Notes
Scratch-out	Starting point	Make minimum three strokes close together
Triangle	Starting point	Don't lift pen while drawing
Square	Starting point	Start at upper-left corner
Star	Starting point	Draw five points
Check	Corner	Draw upward portion 2–4 times larger
Curlicue	Starting point	Angled from lower left to upper right
Double-Curlicue	Starting point	Same as previous but twice

TABLE 18.1 Gestures (*continued*)

Name	Hot Point	Notes
Circle	Starting point	Single stroke starting at top
Double-circle	Starting point	Draw circles overlapping
Left-semicircle	Starting point	Draw from right to left
Right-semicircle	Starting point	Draw from left to right
Caret	Apex	Both sides equal length
Inverted-Caret	Apex	Same as previous but inverted
Chevron-left	Apex	Both sides equal length
Chevron-right	Apex	Both sides equal length
Arrow-up	Arrow head	Limit to two strokes
Arrow-down	Arrow head	Limit to two strokes
Arrow-left	Arrow head	Limit to two strokes
Arrow-right	Arrow head	Limit to two strokes
Up	Starting point	Single upward line (draw fast)
Down	Starting point	Single downward line (draw fast)
Left	Starting point	Single line to left (draw fast)
Right	Starting point	Single line to right (draw fast)
Up-left	Point of change	Single stroke starting with up stroke
Up-right	Point of change	Single stroke starting with up stroke
Down-left	Point of change	Single stroke starting with down stroke
Left-up	Point of change	Single stroke starting with left stroke
Left-down	Point of change	Single stroke starting with left stroke
Right-up	Point of change	Single stroke starting with right stroke
Right-down	Point of change	Single stroke starting with right stroke
Up-down	Point of change	Start with up drawing close together
Down-up	Point of change	Start with down drawing close together
Left-right	Point of change	Start with left drawing close together
Right-left	Point of change	Start with right drawing close together
Up-left-long	Point of change	Start up with left 2x longer

(continues)

TABLE 18.1 Gestures (*continued*)

Name	Hot Point	Notes
Up-right-long	Point of change	Start up with right 2x longer
Down-left-long	Point of change	Start down with left 2x longer
Down-right-long	Point of change	Start up with right 2x longer
Exclamation	Center of line	Draw dot soon after line
Tap	Starting point	Perform quickly
Double-tap	Starting point	Perform quickly

BUILDING THE MEDIA PLAYER

Microsoft provides the Media Player ActiveX control as a simple way to offer media player capabilities. We are going to take advantage of this control as we build an application that can open various movie files and that is controlled by gestures.

To begin this application, create a new Windows Forms application and add an OpenFileDialog to it. You can leave its name as the default given. In the middle of the form, add an InkPicture control, and in the upper right of the form, add a separate one (should look like Figure 18.2) leaving their names as InkPicture1 and InkPicture2.

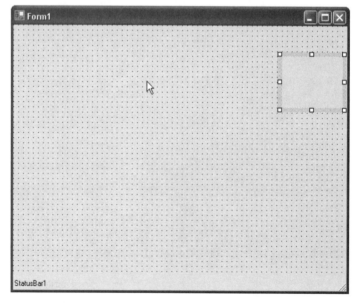

FIGURE 18.2 Position the controls on the form.

Now, let's add the media player control to the form. Choose Project | Add Reference to display the Add Reference dialog box. Next, choose the COM tab and then locate the Windows Media Player OCX control. You can position the media player control on the form as seen in Figure 18.3. This control should approximately cover the InkPlayer control we placed on the form earlier. We're going to use the InkPlayer control to allow us to annotate on the movie.

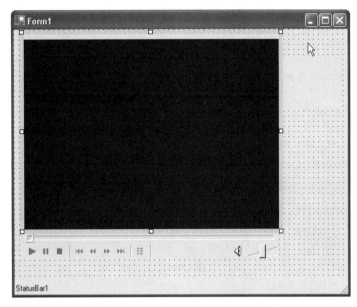

FIGURE 18.3 The Media Player OCX positioned on the form.

The next step for this project is to add a Label control and position it above the InkPicture2 control. Change its Text property to read "Gesture". The final thing we need is a StatusBar control added to the form. When it's added, it automatically positions itself along the bottom of the form, which is where the status bar is normally placed.

Writing Some Code

Before we get to most of the code, we need to add a reference to the Microsoft Ink API. Therefore, we need to add the following line to the top of the code:

```
Imports Microsoft.Ink
```

We'll use the Form_Load event to set several properties very important to our application. First, set the InkPicture1 control to have a transparent background color, remove its border, and then bring it to the front so that it is available for annotating over the video. Here are the three lines of code:

```
InkPicture1.BackColor = Color.Transparent
InkPicture1.BorderStyle = BorderStyle.None
InkPicture1.BringToFront()
```

Now, let's turn our attention to the other InkPicture control. This one will be used to handle the gestures for our application, which will control the features such as opening, playing, and pausing a video. First, we'll set its borderstyle and backcolor so that it separates itself from the rest of the interface and is very visible to the user. As we are interested in using the control to capture gestures only, we are going to set its Collection mode to GestureOnly, and then we'll set the gestures that it will try to recognize. Our application is going to use the following gestures:

Square: Stop

UpDown: Pause

ChevronRight: Play

ArrowRight: Volume increase

ArrowLeft: Volume decrease

Circle: Open

Here is the code:

```
InkPicture2.BorderStyle = BorderStyle.Fixed3D
InkPicture2.BackColor = Color.White
InkPicture2.CollectionMode = CollectionMode.GestureOnly
InkPicture2.SetGestureStatus(ApplicationGesture.Square, True)
InkPicture2.SetGestureStatus(ApplicationGesture.UpDown, True)
InkPicture2.SetGestureStatus(ApplicationGesture.ChevronRight, True)
InkPicture2.SetGestureStatus(ApplicationGesture.ArrowRight, True)
InkPicture2.SetGestureStatus(ApplicationGesture.ArrowLeft, True)
InkPicture2.SetGestureStatus(ApplicationGesture.Circle, True)
```

The last two things we need to handle are the status bar's Text property, which we set to "No file loaded.", and the filter for the open dialog box so that we can only open MPEG and AVI videos:

```
OpenFileDialog1.Filter = "MPEG Video (*.mpg)|*.mpg|Video for Windows
(*.avi)|*.avi|All Files (*.*)|*.*"
StatusBar1.Text = "No file loaded."
```

Now that we have our gestures, we can use the `InkPicture2_Gesture` event to determine which gesture was written and then perform the appropriate actions. We'll use an `If...Then` statement at the beginning of the procedure to check the current gesture's confidence. We can test to see if the recognition confidence is strong, and if so, we will use a `Case` statement to determine which type of gesture was written and instruct the AxMediaPlayer control appropriately. If the gesture confidence is less than strong, we will exit without performing any activities with the AxMediaPlayer. Before we exit, we set the status bar's `Text` property to "`Command Not Recognized`" so that the user realizes that the command they attempted did not perform any function.

We have already discussed the gestures and what each of them will do, so here is the code:

```
If e.Gestures(0).Confidence = RecognitionConfidence.Strong Then
    Select Case e.Gestures(0).Id
        Case ApplicationGesture.Square
            AxMediaPlayer1.Stop()
            StatusBar1.Text = "Stop"
        Case ApplicationGesture.ChevronRight
            AxMediaPlayer1.Play()
            StatusBar1.Text = "Play"
        Case ApplicationGesture.UpDown
            AxMediaPlayer1.Pause()
            StatusBar1.Text = "Pause"
        Case ApplicationGesture.ArrowLeft
            AxMediaPlayer1.Volume = AxMediaPlayer1.Volume - 1
            StatusBar1.Text = "<- Vol = " &
AxMediaPlayer1.Volume.ToString
        Case ApplicationGesture.Circle
            OpenFileDialog1.ShowDialog()
            AxMediaPlayer1.Open(OpenFileDialog1.FileName)
        Case ApplicationGesture.ArrowRight
            AxMediaPlayer1.Volume = AxMediaPlayer1.Volume + 1
            StatusBar1.Text = "-> Vol = " &
AxMediaPlayer1.Volume.ToString
            StatusBar1.Text = AxMediaPlayer1.FileName
    End Select
Else
    StatusBar1.Text = "Command Not Recognized"
End If
```

The application is functional at this time, but we have a few additional things we need to handle. First, we need to change the drawing color for both InkPicture controls, depending on which control the pen is entering. We'll use both `MouseEnter` events for this aspect of the application:

```
Private Sub InkPicture2_MouseEnter(ByVal sender As Object, ByVal e As
System.EventArgs) Handles InkPicture2.MouseEnter
    InkPicture2.DefaultDrawingAttributes.Color = Color.Red
End Sub

Private Sub InkPicture1_MouseEnter(ByVal sender As Object, ByVal e As
System.EventArgs) Handles InkPicture1.MouseEnter
    InkPicture1.DefaultDrawingAttributes.Color = Color.Blue
End Sub
```

The final thing we need to add to the application is a way to pause the AxMediaPlayer currently playing file so that we can annotate a given frame. We can use the `MouseDown` event procedure for InkPicture1 to test the current filename. If there is a filename, we then test to see if the media player is currently playing. If it is, we pause the player so that we can proceed to annotate it. Lastly, we set the status bar's `Text` property to "`Paused for annotation.`"

Here is the code:

```
Private Sub InkPicture1_MouseDown(ByVal sender As Object, ByVal e As
System.Windows.Forms.MouseEventArgs) Handles InkPicture1.MouseDown
    If axMediaPlayer1.FileName  "" and AxMediaPlayer1.PlayState =
MPPlayStateConstants.mpPlaying Then
        AxMediaPlayer1.Pause()
        StatusBar1.Text = "Paused for annotation."
    End If
End Sub
```

You can test the application to see if it works. Figure 18.4 displays an open gesture with an OpenFileDialog and Figure 18.5 shows a video being annotated.

SUMMARY

In this chapter, we created a video player that was controlled using gestures. The application also used an additional InkPicture control to allow us to annotate over the content of the video. In Chapter 19, Getting Started with Microsoft Agent, we build our first speech-enabled application.

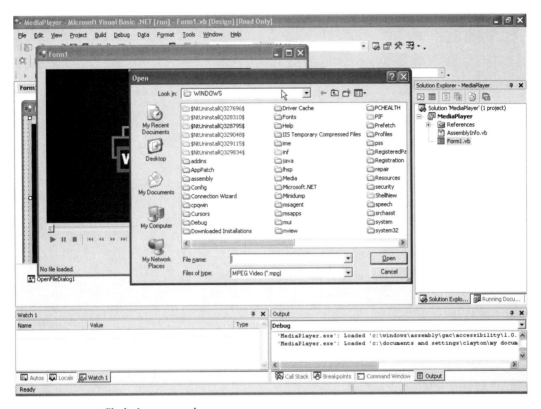

FIGURE 18.4   A file being opened.

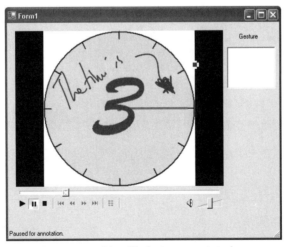

FIGURE 18.5   A video file being annotated.

19 Getting Started with Microsoft Agent

In this chapter, we're again moving away from ink. This time, we look at some additional ways we can communicate with the Tablet PC using the Microsoft Agent ActiveX control that is available freely from Microsoft.

ON THE CD

The source code for the projects are located on the CD-ROM in the PROJECTS folder. You can either type them in as you go or you can copy the projects from the CD-ROM to your hard drive for editing.

WHAT IS AGENT?

Microsoft Agent is a set of tools that can enrich your program. It offers us an easy way to add "animated characters" that can talk by converting Text to Speech (TTS), play animation, and even allow some basic dictation. The "characters" are graphics that are put together in the free Microsoft Agent Character Editor. There are a variety of characters with different capabilities.

When you load a character through Microsoft Agent, its character icon appears in the taskbar. If a mouse is placed over the icon, a ToolTip appears that tells the name of the character, and by single-clicking the icon, it displays the character. We also have the ability to show and hide the character programmatically. The character can be positioned anywhere on the screen by dragging it with the left mouse button. Again, we can also programmatically control this and can position it wherever we want.

When an Agent character talks, a word balloon opens that displays the text version of what the character is saying. When finished talking, the balloon hides automatically. You're probably used to this, but again, we can also control this through code, and we have the ability to hide the balloon during speech.

If a character's speech input is enabled, a user can press the character's push-to-talk button on the keyboard and a listening ToolTip is displayed notifying the user that the character is listening. If the character recognizes something that the user speaks through a microphone, the listening ToolTip displays what the character heard. The character can then react to the commands if we have written code for them. The character will not recognize anything the user speaks unless it was added as a command to the character in the code (more on this in Chapter 20, Advanced Microsoft Agent).

The standard character set includes four unique identities: Peedy the Parrot, the Genie, Merlin the Wizard, and Robby the Robot. In addition to the standard varieties, there are literally hundreds of characters that have been created by third parties. These characters can be simple or very intricate creations, with some of them freely available for download and others commercially available. We'll use Merlin (seen in Figure 19.1) in this chapter, although you'll soon find that it is very easy to switch characters.

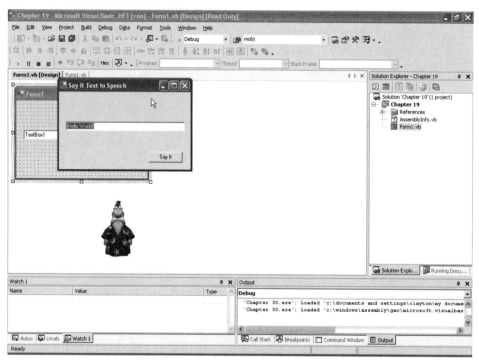

FIGURE 19.1 Merlin, displayed here in the IDE, is one of many characters available for Microsoft Agent.

STARTING THE APPLICATION

Before we begin this application, you should visit the Agent Web site at *http://www.microsoft.com/products/msagent/downloads.htm* to download the Agent control, characters, and Text to Speech engine. All of these are available for free from the Web site, although you may already have some or all of them on your machine.

 We have included Microsoft Agent version 2, which was current at the time of writing, on the CD-ROM that accompanies this book. You can either install it from the CD-ROM, or download it from the previous mentioned Microsoft Agent Web Site when you download the Agent characters and speech engine..

The Microsoft Agent API provides services that support the display and animation of animated characters. Microsoft Agent includes optional support for speech recognition so applications can respond to voice commands (we'll look at this in more detail in Chapter 20, Advanced Microsoft Agent). Characters can respond using synthesized speech, recorded audio, or text in a cartoon word balloon.

Before moving on, you need to download and install the following:

- The Microsoft Agent Core components
- Agent characters Genie, Merlin, Robby, and Peedy
- The Microsoft Speech API 4.0a runtime
- The Microsoft Speech Recognition Engine
- The Lernout and Hauspie Text-to-Speech engines for at least U.S. English

The installations shouldn't be too difficult to follow along with and you might have to go through a series of steps to teach the recognition engine. Make sure to follow the recognition training carefully so that the exercises in Chapter 20 work correctly. After downloading and installing the components and characters, you can open Visual Basic and select Windows Application. Next, right-click on the Standard toolbar and select Customize Toolbox from the pop-up menu (see Figure 19.2).

The Customize Toolbox window is displayed (see Figure 19.3) and contains two tabs. Make sure that the COM Components tab is selected and then scroll the list of available controls until you find Microsoft Agent Control 2.0.

After you find the Agent control, select it from the window and click the OK button. You will now find the Agent control available in the Toolbox among the other controls, although you will probably need to scroll to the bottom of the list. (See Figure 19.4.)

The next step is to add it to the window, just like any standard control. Although it is invisible to the end user, the control is seen at design time and is visible in Figure 19.5.

FIGURE 19.2 Using the pop-up menu.

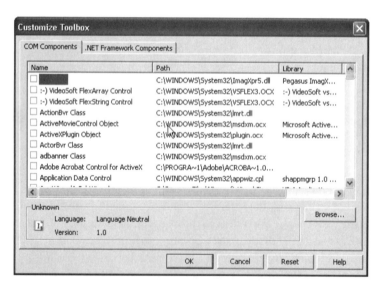

FIGURE 19.3 You can add controls to the Toolbox with the Customize Toolbox window.

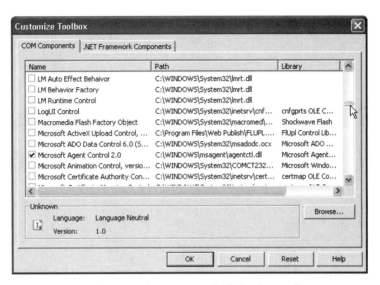

FIGURE 19.4 The control is now available in the Toolbox.

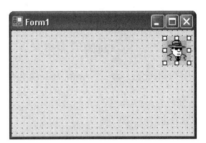

FIGURE 19.5 The control placed on the window.

DECLARING THE CHARACTER FILE

The next step is to open the Code Editor. We are going to declare Agent as type `AgentObjects.IAgentCtlCharacterEx`:

```
Dim Character As AgentObjects.IAgentCtlCharacter
```

We will now set up a variable named sChar that will store the location of the Agent character file named Merlin.acs:

```
Dim sChar As String = "C:\Windows\msagent\chars\merlin.acs"
```

The previous line assumed that the Merlin.acs file was available at a specific location. You can search your local hard drive to locate the file if it is at a different location and change the previous line to reflect the location.

INITIALIZING THE CHARACTER

We are now in a position where we need to initialize the character. We'll do this in the Form_Load event. Before the first step in dealing with the Agent character, we can set the Text property of the form to "Say It Text To Speech". Next, we'll load the Merlin character using the Load method of Agent1 (the control we added to the form). We'll then set an optional property of Character called the LanguageID, which sets the language used for Agent speech. In this case, we'll use English, which has an ID of &H409S.

Here is the code for these lines:

```
        On Error GoTo handler
        Me.Text = "Say It Text to Speech"
        Agent1.Characters.Load("Merlin", sChar)
        Character = Agent1.Characters("Merlin")
        Character.LanguageID = &H409S
handler:
        If Err.Number <> 0 Then MessageBox.Show("Description: " &
Err.Description, "Error!", MessageBoxButtons.OK, _
    MessageBoxIcon.Warning.Warning, MessageBoxDefaultButton.Button1)
        Err.Clear()
```

You may have noticed the simple error handler we include to handle any problems we may encounter with loading the character. We can now add a text box to the form (see Figure 19.6). The text box will be used to enter a string of characters.

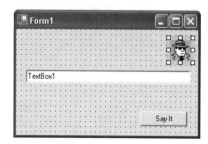

FIGURE 19.6   A text box is added to the form.

The character converts the text information to speech when the user instructs it to do so. In our application, we'll use a Button control for this. Add it to the form and change its Text property as seen in Figure 19.7.

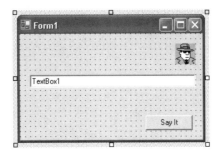

FIGURE 19.7   The Button control is added to the form.

There is a single line of code we need to add to the Form_Load event. This code sets the Text property of TextBox1 to "Hello World!":

```
TextBox1.Text = "Hello World!"
```

The entire procedure is listed in the following code segment:

```
    Private Sub Form1_Load(ByVal eventSender As System.Object, ByVal
eventArgs As System.EventArgs) Handles MyBase.Load
        On Error GoTo handler
        Me.Text = "Say It Text to Speech"
        Agent1.Characters.Load("Merlin", sChar)
        Character = Agent1.Characters("Merlin")
        Character.LanguageID = &H409S
        TextBox1.Text = "Hello World!"
handler:
        If Err.Number <> 0 Then MessageBox.Show("Description: " &
Err.Description, "Error!", MessageBoxButtons.OK, _
    MessageBoxIcon.Warning.Warning, MessageBoxDefaultButton.Button1)
        Err.Clear()
    End Sub
```

We can now concentrate on the Button1_Click event, which will handle the Agent speech and hiding and showing the character. First, we'll use Character.Show to display the agent:

```
Character.Show()
```

Next, we can use the Speak method of the Agent control to get the character to speak. You can use any arrangement of characters, and in our case, we use the `Text` property of TextBox1. The text appears inside a bubble and is also heard:

```
Character.Speak(TextBox1.Text)
```

Once he has spoken, we'll hide the character again until the button is clicked:

```
Character.Hide()
```

The final procedure is as follows:

```
Private Sub Button1_Click(ByVal sender As System.Object, ByVal e As
System.EventArgs) Handles Button1.Click
    Character.Show()
    Character.Speak(TextBox1.Text)
    Character.Hide()
End Sub
```

TESTING THE APPLICATION

At this time, the application is finished and you can save it. To test the application, you can run it in the IDE. On opening, the window should look like Figure 19.8.

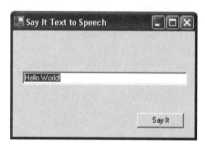

FIGURE 19.8 The screen on opening.

You can click the "Say It" button, which should display the Agent as in Figure 19.9. Next, the Agent speaks the contents of the text box (see Figure 19.10). Lastly, the Agent is hidden and the original window is displayed.

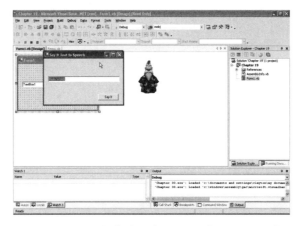

FIGURE 19.9 Displaying the Agent character Merlin.

FIGURE 19.10 The
Agent is speaking.

You can change the contents of the text box and then click the "Say It" button to test the application further.

SUMMARY

In this chapter, we built our first Agent application. With only a few lines of code, we were able to create an application that converted text information into speech. In Chapter 20, Advanced Microsoft Agent, we learn much more about Agent programming and we also build a simple guess the number game.

20 | Advanced Microsoft Agent

I n this chapter, we're going to continue on with the example we started in the previous chapter. In addition to offering the ability to convert text information to speech, Agent also allows us to take speech input, which can be manipulated in a variety of ways.

ON THE CD

The source code for the projects are located on the CD-ROM in the PROJECTS folder. You can either type them in as you go or you can copy the projects from the CD-ROM to your hard drive for editing.

INTRODUCTION TO SPEECH RECOGNITION

If you remember back to Chapter 19, Getting Started with Microsoft Agent, we installed several engines when we installed Microsoft Agent. One of these engines was the speech recognition engine that we use in this chapter. Most speech recognition engines convert incoming audio data to engine-specific phonemes (the smallest structural unit of sound that can be used to distinguish one utterance from another in a spoken language), which are then translated into text that an application can use.

In continuous speech recognition, clients can speak to the system naturally, and the system keeps up with it. On the other hand, discrete recognition requires a user to speak very deliberately and pause between each word. At first glance, it might appear that continuous recognition would always be preferred over discrete recognition. After all, anything you would ever want to do could be accomplished with it. However, continuous speech recognition requires much more processing power, which isn't always available.

If you are planning to develop an application for dictation, you must support a very large vocabulary of words, whereas smaller vocabularies are a satisfactory way to allow a user to send commands to their computers. We use small vocabularies for our applications because we are developing command-based programs.

If you have ever used any type of speech recognition before, you know that you are often required to go through a series of tests to train your system for your voice and speaker. Speaker-independent speech recognition works well with very little or even absolutely zero training, whereas speaker-dependent systems require training, which could amount to hours of your time.

Agent uses "Command and Control" speech recognition, which is continuous, has a small vocabulary, and is speaker independent. With this engine, we can create several hundred different commands or phrases for a program to recognize. If the engine does not recognize a command you give it, the speech-recognition system has two possibilities. The engine could return "not recognized," or it could even mistake it for a command similar to the one you intended. With this in mind, the user of an application must be given the list of phrases they can say, and it is preferable to list them on the screen. You can display the list of commands a given application "listens" for by using the Agent Command window. If speech recognition is disabled, the Voice Commands window is still displayed with the text "Speech input disabled." A character can also have a language setting, so if no speech engine is installed that matches the character's language setting, the window displays, "Speech input not available." If the application has not defined voice parameters for its commands, the window displays, "No voice commands." You can also query the properties of the Voice Commands window (see Figure 20.1) regardless of whether speech input is disabled or a compatible speech engine is installed.

FIGURE 20.1 The Voice
Commands window.

Speech recognition is not enabled at all times, so the end user of an Agent application needs to press a "push-to-talk" key before voice input is enabled. Once enabled, a special ToolTip appears. This listening tip (see Figure 20.2) displays contextual information associated to the current input state.

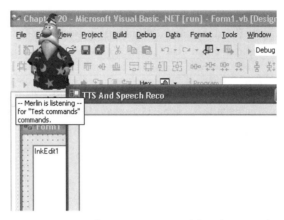

FIGURE 20.2 The current state of the character is visible in the listening tip.

PROGRAMMING BASICS

With the background we have from the previous chapter, it's very easy to make a character recognize a user's speech input. The first thing we need to do is add commands to a character. The commands are the words or phrases that Agent can recognize when a user speaks through his microphone. The character will not recognize any command given by the user unless you program a command for it. To add a command to a character, add the following to your code:

```
Character.Commands.Add("Say Hello.", , "Say Hello.", True, True)
```

Let's break down the previous line into its various pieces so we can understand the "Say Hello" command. The first "Say Hello." is the name of the command name. The command name is how the character "hears" the commands. The second "Say Hello." is the command voice, which is what the listening tip shows. For example, if we had a command name of "Hello" and the command voice was "Hello1," then the listening tip would say "Character heard Hello1" even though the user actually said "Hello." The first of the two "True" statements specifies if the command is enabled, whereas the last one sets the visibility. We're going to define some real commands for our character later in this chapter, but this is an easy-to-follow example so you can see how the commands are broken down.

Adding the command is part of the equation. However, at this time, a character does not do anything if it hears our command. We need to add a procedure to make a character respond to the "Say Hello." command. We can use the

Agent_Command Sub and an If...Then statement to determine what the character says and how it will respond. Continuing with our "Say Hello." example, the following procedure allows the character to say hello:

```
Private Sub Agent1_Command(ByVal sender As Object, ByVal e As
AxAgentObjects._AgentEvents_CommandEvent) Handles Agent1.Command
    Dim command As AgentObjects.IAgentCtlUserInput = CType(e.userInput,
AgentObjects.IAgentCtlUserInput)

        If command.Name = "Say Hello." Then
            Character.Speak("Hello")
        End If

End Sub
```

Advanced Commands

Although the "Say Hello" command we used to demonstrate the speech recognition process is very simple, the commands we use often become much more complicated. When we define the voice grammar, we can use punctuation and symbols to enhance our offerings. These options include brackets ([ ]), stars (*), addition signs (+), parentheses ("("and ")"), and vertical bars (|).

The following list details the grammar options:

"*" and "+": You use the star to specify zero or more instances of a word. For example, if you use a command such as "Please* open file," the character recognizes it regardless of how many times "Please" was said. That is, if "Open file" is said, "Please open file" is recognized. Likewise, "Please Please open file" is also recognized as "Please open file." You use the plus sign in the same way, but the plus sign requires one or more instances of the word.

"[ ]": You can use brackets to indicate optional words. This is similar to the "*" command. For example, we can use "[Please] open file" like we did "Please* open file." Please is optional in either example.

"( )": Parentheses are used to indicate alternative words. As an example, you can use "Please open (the) file," which allows "the" to be optional for the command.

"|": You use the vertical bar to let the character know that the word can be pronounced in different ways. As an example, some people may pronounce the word "the" differently. That is, someone could pronounce it as "thee," whereas others pronounce it as "the." You can take care of this problem using the vertical bar as follows: "Open (thee|the) file." This example combined two of the options to allow the user to say "Open thee file," "Open the file," or "Open file" and all are recognized by the application. You could also use numbers with the vertical bar: "Please call(five|five|five|one|two|one|two)" allows the person to say "Please call 555-1212."

Reacting to Other Events

Agent can also respond to other events such as when a user clicks on the character. The click action causes Agent to fire an event that passes back the button that was clicked, any modifier keys that were pressed, and the x and y coordinates of the mouse. An example event handler for this is as follows:

```
Sub Agent_Click(ByVal CharacterID As String, ByVal Button As Integer,
ByVal Shift As Integer, ByVal X As Integer, ByVal Y As Integer)

    Agent1.Speak "Click event"

End Sub
```

Agent also includes support for a context menu, when the user right-clicks the character.

REAL-WORLD USEFULNESS

Now that we have seen how to add commands to Agent, we will use them in our example. Open the file from the previous chapter and add two buttons with the properties shown in Table 20.1.

TABLE 20.1   Adding btnProperties and btnListen

Name	Text
btnProperties	Properties
btnListen	Listen

You can also replace the TextBox1 control with an InkEdit control. You can see the buttons added to the form and the replaced controls in Figure 20.3.

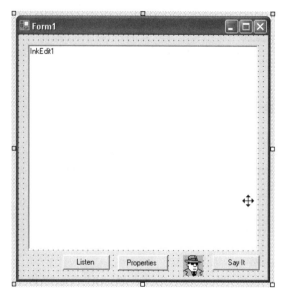

FIGURE 20.3 The GUI is complete.

Next, in the Code Editor, change any references for TextBox1 to InkEdit1, and remove any references that set a character to show and hide because we want the character to be visible at all times. In addition, remove the "Hello World!" text from initializing InkEdit1 as follows:

```
Private Sub Form1_Load(ByVal eventSender As System.Object, ByVal
eventArgs As System.EventArgs) Handles MyBase.Load
    On Error GoTo handler
    Agent1.Characters.Load("Merlin", sChar)
    Character = Agent1.Characters("Merlin")
    Character.LanguageID = &H409S
    InkEdit1.Text = ""
handler:
    If Err.Number <> O Then
MessageBox.Show("Description: " & Err.Description, "Error!",
MessageBoxButtons.OK, _
MessageBoxIcon.Warning.Warning, MessageBoxDefaultButton.Button1)
    Err.Clear()
End Sub

Private Sub Button1_Click(ByVal sender As System.Object, ByVal e As
System.EventArgs) Handles Button1.Click
    Character.Speak(InkEdit1.Text)
End Sub
```

Next, we can add the commands to our character using the `Form_Load` event. For starters, let's add a caption for our commands (the lines can be added before the error handler and directly beneath initializing InkEdit1 with an empty string):

```
Character.Commands.Caption = "Test commands"
```

The next step is to add several commands to our character. We're going to instruct the character to read the clipboard, say the contents of InkEdit, and close all via speech input. Here is the code:

```
Character.Commands.Add("Read Clipboard", "Read Clipboard", True, True)
Character.Commands.Add("Say Ink", "Say Ink", True, True)
Character.Commands.Add("exit|close|quit", "exit", True, True)
Character.Show()
```

We now need to create the event to respond to the Agent's commands. We use a `Case` statement to look at the possible choices and respond appropriately. Here is the code for the procedure:

```
Private Sub Agent1_Command(ByVal sender As Object, ByVal e As
AxAgentObjects._AgentEvents_CommandEvent) Handles Agent1.Command
    Dim command As AgentObjects.IAgentCtlUserInput = CType(e.userInput,
AgentObjects.IAgentCtlUserInput)

    Select Case command.Name
        Case "exit"
            End
        Case "Say Ink"
            Button1.PerformClick()
        Case "Read Clipboard"
            Character.Speak(Clipboard.GetDataObject())
    End Select

End Sub
```

By default, the Scroll Lock key instructs the Agent character to listen to commands. Although this is fine for many situations, a Tablet PC user may or may not have access to a keyboard. Therefore, we provide two ways to cause the character to listen for keys.

First, we create a click event for btnListen. This procedure begins by stopping anything that the character is currently doing. Next, we set the character to listen:

```
Private Sub btnListen_Click(ByVal sender As System.Object, ByVal e As
System.EventArgs) Handles btnListen.Click
```

```
      Character.StopAll()
      Character.Listen(True)
   End Sub
```

We've already mentioned the Tablet PC user may or may not have access to their keyboard. At this time, the user could click btnListen to start listening for any voice commands. This works similarly to the push-to-talk Listening hotkey, but if the user prefers to press one of the buttons they have available (such as the Tab key, which is included on the HP 1000 Tablet PC when in slate mode), we should provide this option. Unfortunately, we cannot set the push-to-talk Listening hotkey programmatically in our application, but we do have access to the Agent's Property sheet (see Figure 20.4), which allows the user to set the key themselves:

```
Private Sub btnAgentProperties_Click(ByVal sender As System.Object,
ByVal e As System.EventArgs) Handles btnAgentProperties.Click
    Agent1.PropertySheet.Visible = True
End Sub
```

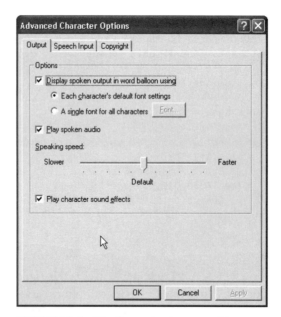

FIGURE 20.4 The Property sheet provides options for the end user.

You can now test the application to make sure that it functions correctly. You should spend the time to test both the voice input (see Figure 20.5) and speech output (see Figure 20.6), along with changes to the Agent Property sheet.

FIGURE 20.5   Listening for a command.

FIGURE 20.6   Reading content of InkEdit Control.

SUMMARY

In this chapter, we built our second application with Microsoft Agent. In this example, we used Agent and its speech recognition engine to allow us to give commands verbally. In Chapter 21, Speech Input with SAPI, we build another speech recognition program, but it will be our first with the Speech SDK version 5.1.

21 Speech Input with SAPI

In the previous chapter, we built our second application with Microsoft Agent. The application took advantage of speech recognition technologies available as an engine for the Agent characters. In this chapter, we continue on a similar path, but this time, we use the Microsoft Speech SDK.

The source code for the projects are located on the CD-ROM in the PROJECTS folder. You can either type them in as you go or you can copy the projects from the CD-ROM to your hard drive for editing.

5.1 SDK OVERVIEW

When the Microsoft Research Speech Technology Group released the SAPI 5.0 SDK, it contained one big omission, a COM interface for VB programmers. By the time version 5.1 was released, they had fixed this shortcoming. It's important not to confuse the SAPI 5.1 SDK with the .NET Speech SDK. The .NET Speech SDK should only be used for Web-based applications (Web Forms), whereas the SAPI SDK was intended for Windows Forms applications.

The SAPI speech recognition engine differs greatly from the Agent offering. For starters, the engine is advanced enough to allow you to dictate directly into an application. You could dictate letters or memos without the need to make many corrections afterwards. The engine even works very well without extensive training. You can download the Speech SDK from *http://www.microsoft.com/speech/download/sdk51/*, and remember that we need to use the version 5.1 Speech SDK in lieu of the .NET Speech SDK because it is for Windows-based applications rather than the Web-centric .NET version.

Like Agent, the SAPI speech recognition engine is also very easy to use. We can begin the application by creating a new Windows Forms application. The next step is to add a reference to the Microsoft Speech Object Library. You can add the library using the Project menu, selecting Add Reference, clicking on the COM tab, and then choosing the library from the list.

The user interface for our application will be very simple and will consist of four buttons, a status bar, and a text box. You can use the properties listed in Table 21.1 and refer to Figure 21.1 for the arrangement of the controls.

TABLE 21.1   Adding buttons and the status bar to our user interface

Type	*Name*	*Text*
Button	btnSpeak	Speak
Button	btnSpeaktoFile	Speak To File
Button	btnRecognize	Recognize
Button	btnStopRecognition	Stop Recognition
StatusBar	StatusBar1	StatusBar1
TextBox	txtSpeech	TextBox1

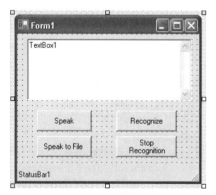

FIGURE 21.1   The completed GUI for our application.

You could replace the standard text box with one of the Inkable controls if you want to make it ink- and speech-enabled.

WRITING SOME CODE

With the brief overview and GUI out of the way, we can write some code for our application. We begin the code with the `Imports` statements for `SpeechLib` and `System.Threading`:

```
Imports SpeechLib
Imports System.Threading
```

The `SpeechLib` was added as a reference earlier in the project and probably comes as no surprise. The `System.Threading` namespace is something that we have yet to touch on. In simple terms, threading is the ability to run different pieces of code (methods) at the same time. You can even think of it as multitasking within a single application. Although that sounds good, a computer can't do multiple things at the same time. Therefore, in order for threading to work, different tasks and threads have to share processing resources. It is the job of the operating system to assign the time to the different tasks. As a real-world example, take a look at Microsoft Internet Explorer. As you navigate a Web site, items are being downloaded. As busy as the application is retrieving information, you can still perform other tasks, such as move the window around, resize it, and even open another window. This is because all of the tasks are running on different threads.

We don't have much of a need in the way of threads for this application. Really, we only need to use a constant called `Timeout.Infinite`, which is used to specify an infinite waiting period. We use the constant in this application when we save the text-to-speech output of our application to a WAV file.

Our next step is to `Dim` some variables that we'll need to use throughout the application. These variables include `RecoContext`, `Grammar`, `m_BrecoRunning`, and `m_cChars`.

```
Dim WithEvents RecoContext As SpeechLib.SpSharedRecoContext
Dim Grammar As SpeechLib.ISpeechRecoGrammar
Dim m_bRecoRunning As Boolean
Dim m_cChars As Short
```

We use the `Form_Load` event to initialize some of the controls and variables. First, we set `txtSpeech` (or a replacement Ink control) to an empty string value. This effectively removes any text it currently has, so that anything we dictate into the application is visible by itself. Next, we set the state of the recognition to `False`. We'll actually create a `Sub` routine for this, but for now, you can simply type the line that calls the routing. Additionally, we set `m_cChars` equal to `0` and the `Text` property of StatusBar1 equal to an empty string.

Here is the code:

```
txtSpeech.Text = ""
SetState(False)
m_cChars = 0
StatusBar1.Text = ""
```

The next step is to create the SetState Sub routine. This Sub allows us to turn the recognition on and off quickly and effectively throughout the use of the application. We'll begin by passing a Boolean value into the Sub, because the True and False values are an easy way to set the current state of the recognition. If you pass a True value, recognition is working, whereas False obviously turns it off.

The value being passed in will be known as bNewState by the Sub. We can use bNewState directly to set the m_bRecoRunning variable. After setting the value, we can set the enabled property of btnRecognize equal to the opposite of m_bRecoRunning. Therefore, if the engine is running, the button is not enabled. Additionally, if the engine is not running, the button is enabled. The last step in the Sub is to use m_bRecoRunning to set the enabled property of btnStopRecognize. This time, however, we can use the actual value as we only want btnStopRecognize enabled when the engine is running and we certainly don't have a need to click on it when it is not running.

Here is the entire procedure:

```
Private Sub SetState(ByVal bNewState As Boolean)
    m_bRecoRunning = bNewState
    btnRecognize.Enabled = Not m_bRecoRunning
    btnStopRecognize.Enabled = m_bRecoRunning
End Sub
```

Let's now look at the events that are triggered when each of the four buttons are clicked as well as the RecoContext event, which occurs when m_bRecoRunning is enabled and some speech is recognized. We'll start with the btnRecognize_click event.

The btnRecognize_click event begins with initializing the Recognition Context object and the Grammar object. We will then use the SetStatus Sub procedure, and by passing a value of True, we set the state of the recognition to True. Lastly, we set the Text property of StatusBar1 to "SAPI ready for dictation...".

```
Private Sub btnRecognize_Click(ByVal sender As System.Object, ByVal e
As System.EventArgs) Handles btnRecognize.Click
    If (RecoContext Is Nothing) Then
        StatusBar1.Text = "Initializing SAPI reco..."
        RecoContext = New SpeechLib.SpSharedRecoContext()
```

```
        Grammar = RecoContext.CreateGrammar(1)
        Grammar.DictationLoad()
    End If

    Grammar.DictationSetState(SpeechLib.SpeechRuleState.SGDSActive)
    SetState(True)
    StatusBar1.Text = "SAPI ready for dictation..."
End Sub
```

This is the first time in our application that the state of the recognition has been set to True, so it seems like an appropriate time to look at the `RecoContext` `Recognition` method. You can create this procedure by using the Class Name and Method Name drop-down lists and choosing RecoContext and Recognition, respectively.

We begin by setting the StatusBar1 `Text` property to `"Recognizing..."`. This gives an update to our user that something is actually occurring in our application. If you did not update the user on the status, it might appear to them that the application was not working correctly. Most of the remaining portion of code for this procedure is used to add the recognized text to the text box. When we append the text to the text box, we add a space so that the sentences do not run together. Once finished, we set the StatusBar1 `Text` property to `"Speech recognized successfully...SAPI enabled"` so the user knows he can continue dictation if he wants.

Here is the code:

```
Private Sub RecoContext_Recognition(ByVal StreamNumber As Integer,
ByVal StreamPosition As Object, ByVal RecognitionType As
SpeechLib.SpeechRecognitionType, ByVal Result As
SpeechLib.ISpeechRecoResult) Handles RecoContext.Recognition
    Dim strText As String
    strText = Result.PhraseInfo.GetText
    StatusBar1.Text = "Recognizing..."
    txtSpeech.SelectionStart = m_cChars
    txtSpeech.SelectedText = strText & " "
    m_cChars = m_cChars + 1 + Len(strText)
    StatusBar1.Text = "Speech recognized successfully...SAPI enabled"
End Sub
```

Now that we are dictating into the application, we also need a way to turn the dictation process off. This job is taken care of with btnStopRecognize. When we click the button, we want to set the grammar state to inactive, use `SetState` to set the recognition to `False`, and then update the StatusBar1 to reflect these changes.

Here is the code:

```
Private Sub btnStopRecognize_Click(ByVal sender As System.Object, ByVal
e As System.EventArgs) Handles btnStopRecognize.Click
    Grammar.DictationSetState(SpeechLib.SpeechRuleState.SGDSInactive)
    SetState(False)
    StatusBar1.Text = "SAPI Disabled"
End Sub
```

Dictation into the application is now taken care of completely, although we have a few additional features we want to add to our application. In addition to speech recognition, SAPI allows us to perform Text to Speech similarly to Microsoft Agent. We have two remaining buttons on our form, and their click events allow us to read back the text audibly, or, if we choose, we can save the content of the text box to a WAV file so that it could be played later.

Let's handle btnSpeak first because it is the less complicated of the two and covers part of what occurs in btnSpeakToFile. We begin by creating a variable called Voice and setting it to SpVoice. We can then use the Speak method of Voice to convert the text contents of the txtSpeech text box to speech. Lastly, we update Status-Bar1 to "SAPI disabled...".

At this time, we are not enabling and disabling the Speak and Speak to File buttons. If you plan to add anything to this application or distribute it to an end user, you can take care of this within the SetState *Sub routine.*

Here is the code:

```
Private Sub btnSpeak_Click(ByVal sender As System.Object, ByVal e As
System.EventArgs) Handles btnSpeak.Click
    Dim Voice As SpVoice
    Voice = New SpVoice()
    Voice.Speak(txtSpeech.Text,
SpeechLib.SpeechVoiceSpeakFlags.SVSFlagsAsync)
StatusBar1.Text = "SAPI disabled..."
End Sub
```

Like btnSpeak, btnSpeakToFile begins with initializing the Voice variable. Next, we set the Text property of StatusBar1 to "Saving to file...". We then proceed to create a new SaveFileDialog and create a filter for saving the contents as a WAV file. Once we have a filename, we then Dim spFileMode and spFileStream so that we know how to save the file. We then use the Speak method of Voice, but instead of audibly hearing the speech, it is saved into the WAV file we create. Lastly, we close the stream and then set the StatusBar1 Text property to "SAPI disabled...".

Here is the code:

```
Private Sub btnSpeakToFile_Click(ByVal sender As System.Object, ByVal e
As System.EventArgs) Handles btnSpeakToFile.Click
    Dim Voice As SpVoice
    Voice = New SpVoice()
    StatusBar1.Text = "Saving to file..."

    Dim sfd As SaveFileDialog = New SaveFileDialog()

    sfd.Filter = "All files (*.*)|*.*|wav files (*.wav)|*.wav"
    sfd.Title = "Save to a wave file"
    sfd.FilterIndex = 2
    sfd.RestoreDirectory = True

    If sfd.ShowDialog() = DialogResult.OK Then
        Dim SpFileMode As SpeechStreamFileMode =
SpeechStreamFileMode.SSFMCreateForWrite
        Dim SpFileStream As SpFileStream = New SpFileStream()

        SpFileStream.Open(sfd.FileName, SpFileMode, False)

        Voice.AudioOutputStream = SpFileStream
        Voice.Speak(txtSpeech.Text,
SpeechLib.SpeechVoiceSpeakFlags.SVSFlagsAsync)
        Voice.WaitUntilDone(Timeout.Infinite)

        SpFileStream.Close()
    End If
    StatusBar1.Text = "SAPI Disabled..."
End Sub
```

TESTING THE APPLICATION

You are now in position to test the application after you save it. When you start it, the screen should look like Figure 21.2.

You can now try the various buttons to see how each of them performs. When you test the Speak to File button, you can test the resulting WAV file by double-clicking it. The application that has been set to open WAV files (by default, this is Media Player) opens and plays the file similarly to Figure 21.3.

FIGURE 21.2   Your application should be similar on startup.

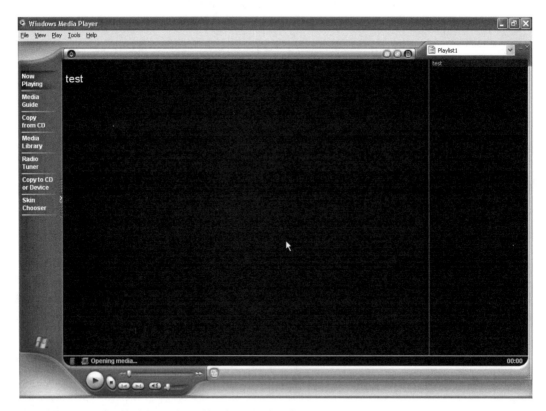

FIGURE 21.3   The file being played back in Media Player.

SUMMARY

In this chapter, we built the first of two programs based on the SAPI 5.1 Speech SDK. We used the built-in grammar as we tested the speech recognition capabilities. Although it is certainly very good, there are certain types of applications that can benefit from more precise recognition. This brings us to Chapter 22, Custom Grammars for Speech Recognition, in which we build an application that performs basic arithmetic. The numbers and operations are all completed by speech recognition using a custom grammar.

22 Custom Grammars for Speech Recognition

In the previous chapter, we built an application using the SAPI 5.1 SDK. With very little voice training, an end user could use the program for basic dictation and text-to-speech output. Although the general grammar used by the application works well for purposes like dictation, a very specific grammar can be beneficial in particular situations.

In this chapter, we build an application that allows number and mathematical operations to be entered by speech. The program then computes the answer automatically.

ON THE CD

The source code for the projects are located on the CD-ROM in the PROJECTS folder. You can either type them in as you go or you can copy the projects from the CD-ROM to your hard drive for editing.

CREATING A CUSTOM GRAMMAR

Much of the programming in this application is similar to the previous SAPI example. Therefore, we begin by creating our custom grammar. The grammar rules used by SAPI are defined using XML (eXtensible Markup Language). This is very attractive to those with an HTML or any XML-derivative background and makes writing grammar fairly easy.

We use Notepad, although you could use any text editor, XML editor, HTML editor, or even Visual Studio if you prefer. Let's begin by opening Notepad. Add the following line to the empty document:

```
<?xml version="1.0" encoding="utf-8" ?>
```

The grammar itself is surrounded by "GRAMMAR" tags. The next line consists of the rules, the first (and only for our application) of which is the number rule, which has an "ACTIVE" tag associated with it, meaning this is something the speech recognition engine should use.

Add the following lines to your code:

```
<GRAMMAR>
<RULE ID="1" Name="number" TOPLEVEL="ACTIVE">
</GRAMMAR>
```

At this time, your Notepad document should contain the following lines:

```
<?xml version="1.0" encoding="utf-8" ?>
<GRAMMAR>
<RULE ID="1" Name="number" TOPLEVEL="ACTIVE">
</GRAMMAR>
```

Before moving on, we need to take a quick look at the various elements we'll encounter:

<L>: Defines an expression of alternate phrase recognitions. Each subelement represents a possible separate recognition in place of this element. It is a synonym of the LIST tag. Empty elements are not valid (i.e., the tag must have children). The LIST element can define a default property name (PROPNAME) or ID (PROPID), which is inherited by its child PHRASE elements.

<P>: Describes the PHRASE element. It is a synonym of the PHRASE element. An associated property name and value pair is generated only if the contents of this element are recognized. It is important to note that a P empty element is not allowed.

We also need to understand the grammar attributes:

<...VALSTR="">: Specifies the string value to be associated with the semantic property (name/value pair)

<...PROPNAME="">: Specifies the string identifier to be associated with the semantic property (name/value pair)

<...VAL="">: Specifies the numeric value to be associated with the semantic property (name/value pair)

Now that we have the attributes and elements to work with, it's very easy to fill in the remaining part of the grammar. We need to recognize the following operations: plus, minus, times, divided by, quit, equal, and new.

Here are the operations:

```
<P VALSTR="PLUS">plus</P>
<P VALSTR="MINUS">minus</P>
<P VALSTR="TIMES">times</P>
<P VALSTR="DIVIDED BY">divided by</P>
<P VALSTR="QUIT">QUIT</P>
<P VALSTR="EQUAL">EQUAL</P>
<P VALSTR="NEW">NEW</P>
```

The last part of the XML file should handle all of the numeric entries from 0 to 30. Here are the entries:

```
<P VAL="0">zero</P>
<P VAL="1">one</P>
<P VAL="2">two</P>
<P VAL="3">three</P>
<P VAL="4">four</P>
<P VAL="5">five</P>
<P VAL="6">six</P>
<P VAL="7">seven</P>
<P VAL="8">eight</P>
<P VAL="9">nine</P>
<P VAL="10">ten</P>
<P VAL="11">eleven</P>
<P VAL="12">twelve</P>
<P VAL="13">thirteen</P>
<P VAL="14">fourteen</P>
<P VAL="15">fifteen</P>
<P VAL="16">sixteen</P>
<P VAL="17">seventeen</P>
<P VAL="18">eighteen</P>
<P VAL="19">nineteen</P>
<P VAL="20">twenty</P>
<P VAL="21">twenty-one</P>
<P VAL="22">twenty-two</P>
<P VAL="23">twenty-three</P>
<P VAL="24">twenty-four</P>
<P VAL="25">twenty-five</P>
<P VAL="26">twenty-six</P>
<P VAL="27">twenty-seven</P>
<P VAL="28">twenty-eight</P>
<P VAL="29">twenty-nine</P>
<P VAL="30">thirty</P>
```

The application will only recognize the values that you place into the file. As such, if you need to place, for example, something like "45," you need to add the values up to 45 to the file.

The complete XML file should contain all of these lines as follows:

```
<?xml version="1.0" encoding="utf-8" ?>
<GRAMMAR>
<RULE ID="1" Name="number" TOPLEVEL="ACTIVE">
<L PROPNAME="number">
<P VALSTR="PLUS">plus</P>
<P VALSTR="MINUS">minus</P>
<P VALSTR="TIMES">times</P>
<P VALSTR="DIVIDED BY">divided by</P>
<P VALSTR="QUIT">QUIT</P>
<P VALSTR="EQUAL">EQUAL</P>
<P VALSTR="NEW">NEW</P>
<P VAL="0">zero</P>
<P VALSTR="1">one</P>
<P VAL="2">two</P>
<P VAL="3">three</P>
<P VAL="4">four</P>
<P VAL="5">five</P>
<P VAL="6">six</P>
<P VAL="7">seven</P>
<P VAL="8">eight</P>
<P VAL="9">nine</P>
<P VAL="10">ten</P>
<P VAL="11">eleven</P>
<P VAL="12">twelve</P>
<P VAL="13">thirteen</P>
<P VAL="14">fourteen</P>
<P VAL="15">fifteen</P>
<P VAL="16">sixteen</P>
<P VAL="17">seventeen</P>
<P VAL="18">eighteen</P>
<P VAL="19">nineteen</P>
<P VAL="20">twenty</P>
<P VAL="21">twenty-one</P>
<P VAL="22">twenty-two</P>
<P VAL="23">twenty-three</P>
<P VAL="24">twenty-four</P>
<P VAL="25">twenty-five</P>
<P VAL="26">twenty-six</P>
<P VAL="27">twenty-seven</P>
<P VAL="28">twenty-eight</P>
<P VAL="29">twenty-nine</P>
```

```
<P VAL="30">thirty</P>
</L>
</RULE>
</GRAMMAR>
```

After you have entered all of the text into Notepad, you can save the XML file. To save a file in Notepad with an extension other than ".txt," you must choose Save As from the File menu and then use quotation marks around the filename. In our case, we need to save the file as "grammar.xml." You can save this to your desktop or some other place that is easily accessible. Later, we'll copy this file to the "bin" directory of our application so that is available when we run the program.

USER INTERFACE

The custom grammar is now finished and is probably the most important thing we are going to create. However, in order to test the XML-based grammar, we need to build an application that loads it. We begin with a GUI that consists of a few controls, shown in Figure 22.1. You can use the figure as a guide to add the controls found in Table 22.1 to the form:

TABLE 22.1   Adding controls to the GUI

Type	Name	Text
TextBox	txtSpeech	TextBox1
Label	lblFirst	0
Label	lblOperand	+
Label	lblSecond	0
Label	lblAnswer	=

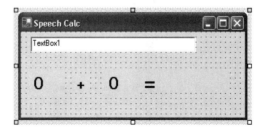

FIGURE 22.1   The finished GUI.

You already know that this application differs from the previous example because we are going to load a custom grammar. This is the biggest change, but it is
definitely not the only one. One of the changes is in the way that recognition is handled. Rather than clicking a button to start the recognition, the application is
speech-ready on startup. All input for this application is speech-enabled. That is,
you never need a mouse or your pen to do anything. You have the ability to close
the application and control all input using only speech.

LOAD GRAMMAR

We begin the programming part of the application by adding the reference to the
Microsoft Speech Object Library and then adding the Imports statement as we did
in the previous example. We also create the same variables, although we don't have
a need for m_bRecoRunning because the recognition engine is always running.

Here are the three Dim statements:

```
Dim WithEvents RecoContext As SpeechLib.SpSharedRecoContext
Dim Grammar As SpeechLib.ISpeechRecoGrammar
Dim m_cChars As Short
```

The Form_Load event will be used for initializing several variables and loading
the grammar. Most of the code, with the exception of loading the grammar, should
look very similar to the previous example. As you can see from the following code,
we are loading our custom Grammar.xml file. This is a good time to copy the
Grammar.xml file from the location you saved it to earlier, to the "bin" folder for
our project. Without this file, you simply receive an error message and the application does not run.

Here is the code for the procedure:

```
txtSpeech.Text = ""
m_cChars = 0
lblFirst.Text = ""
lblSecond.Text = ""
If (RecoContext Is Nothing) Then
    RecoContext = New SpeechLib.SpSharedRecoContext()
    Grammar = RecoContext.CreateGrammar(1)

Grammar.CmdLoadFromFile(System.AppDomain.CurrentDomain.BaseDirectory()
& "grammar.xml", SpeechLoadOption.SLOStatic)
```

```
        Grammar.DictationSetState(SpeechRuleState.SGDSInactive)
        Grammar.CmdSetRuleIdState(1, SpeechRuleState.SGDSActive)
    End If
```

Recognition

The recognition event for this application is handled in exactly the same way the earlier application was handled. We use a Case statement to determine the recognized text and then perform the appropriate changes to the labels and text box. The application works as follows:

1. The application is opened and everything is blank.
2. Recognition starts for the first number (lblFirst).
3. Recognition occurs for the operand (lblOperand).
4. Recognition for the last number takes place (lblSecond).
5. "Equal" is said by the user to perform the calculation (lblAnswer).
6. The user has several options. He can continue to dictate a different operand or second number and obtain different answers when doing so by saying "Equal." He can also say "New" or "Quit" to start a new problem or exit the application. If he starts a new equation, the project starts back at step 2.

Here is the code for the procedure:

```
Dim strText As String
strText = Result.PhraseInfo.GetText

Select Case strText
    Case "plus"
        lblOperand.Text = "+"
    Case "minus"
        lblOperand.Text = "-"
    Case "divided by"
        lblOperand.Text = "/"
    Case "times"
        lblOperand.Text = "*"
    Case "QUIT"
        End
    Case "EQUAL"
        If lblFirst.Text <> "" And lblSecond.Text <> "" Then
            Dim X, Y As Integer
            X = Int32.Parse(lblFirst.Text)
            Y = Int32.Parse(lblSecond.Text)
```

```
            If lblOperand.Text = "+" Then
                lblAnswer.Text = "= " & (X + Y).ToString
            ElseIf lblOperand.Text = "-" Then
                lblAnswer.Text = "= " & (X - Y).ToString
            ElseIf lblOperand.Text = "/" Then
                lblAnswer.Text = "= " & (X / Y).ToString
            ElseIf lblOperand.Text = "*" Then
                lblAnswer.Text = "= " & (X * Y).ToString
            End If
        End If
    Case "NEW"
        lblFirst.Text = ""
        lblSecond.Text = ""
        lblAnswer.Text = "="
    Case Else
        If lblFirst.Text = "" Then
            lblFirst.Text = Result.PhraseInfo.Properties.Item(0).Value
        Else
            If lblSecond.Text = "" Then
                lblSecond.Text =
Result.PhraseInfo.Properties.Item(0).Value
            End If
        End If
End Select

txtSpeech.Text = Result.PhraseInfo.Properties.Item(0).Value
```

TESTING THE APPLICATION

Testing this application is very simple. When you run the application, it looks like Figure 22.2. Next, say a number, such as "Three." You should now see the application recognize the number in the text box, and it also changes the first of the two labels to the same number (see Figure 22.3). Next, give the application the operation you want to perform. For example, you can say "Times" to perform multiplication (see Figure 22.4). The last number is entered in the same way as the first, so if you say "Five," it is displayed in the text box and changes the second label (see Figure 22.5). Finally, you can say "Equal" to perform the calculation (see Figure 22.6).

If you want to continue performing calculations, you can say "New" to start over.

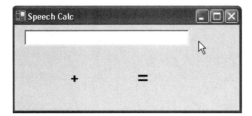

FIGURE 22.2   The application on startup.

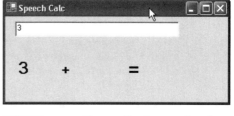

FIGURE 22.3   The application receives its first digit.

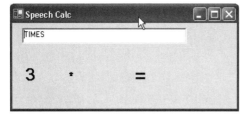

FIGURE 22.4   The operand is recognized.

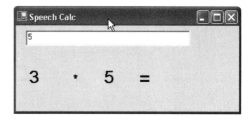

FIGURE 22.5   The final number is recognized.

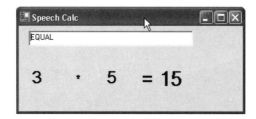

FIGURE 22.6   The calculation is complete.

SUMMARY

In this chapter, we created an application that essentially uses speech recognition to perform basic calculations. We created a custom grammar, so the application only recognizes the items we want it to. You can see how the grammars can be a very effective method to increase speech recognition accuracy for certain types of applications. In the next two chapters, we begin our look at the hardware of the Tablet PC and how we can develop software around it.

23

WMI and Hardware

In this chapter, we develop a program that is based on the WMI class, or the Windows Management Instrumentation. WMI is an industry initiative to develop a standardized technology for accessing management information in enterprise environments. This information includes the state of system memory, inventories of currently installed client applications, and various other pieces of data (see Figure 23.1). Hardware information, especially when dealing with devices that are often on battery power like a Tablet PC, can be important to a developer. This chapter goes hand in hand with Chapter 24, Power Management for the Tablet PC, which details the power management capabilities that we can access.

FIGURE 23.1 Our finished application with information displayed.

The source code for the projects are located on the CD-ROM in the PROJECTS folder. You can either type them in as you go or you can copy the projects from the CD-ROM to your hard drive for editing.

PROJECT OVERVIEW

In this project, we take advantage of the WMI classes that are included in VB .NET to build an application that lists the various components and information about your PC. Our application will include the following information, but as you'll see, there is much more that can be included by making a few simple changes to this program.

Here is the information we're going to display:

- Network card
    - MAC address of network card
    - Network card description
- BIOS information
    - Name
    - Serial number
    - Manufacturer
    - Release date
    - SMBIOS version
    - SMBIOS major version
    - SMBIOS minor version
    - Software element ID
    - Software element state
    - Version
    - Current language
- Computer system information
    - Caption
    - Primary owner name
    - Domain
    - Domain role
    - Manufacturer
    - Number of processors
    - System type
    - System startup display
    - Total physical memory

- Processor information
    - CPU name
    - Voltage caps
    - L2Cache
    - Current clock speed
    - CPU status
- Pointing device
    - Device ID
    - Pointing type
    - Manufacturer
    - Number of buttons
    - Status

You can see that this program would be quite useful for troubleshooting information or simply to retrieve information from a system. This information could be used for copy protection code by retrieving a serial number and then writing a key generation program to generate a single valid key based on this number. You could also use this in an About Box to give the user details about their system. Additionally, you could use this to help determine if a keyboard or external mouse was connected to a Tablet PC. Suffice it to say, there are many ways in which you could put this information to good use.

GETTING STARTED

This application will consist of a very simple user interface. We could just as easily have used a Console application template for this project, although we use a Windows Form template. This allows us much more flexibility if we want to add the ability to e-mail this information to someone or want to expand this program at a later date to include more information. Additionally, we could simply use this in another application as an About Box or a troubleshooting screen.

This first step is to create a Windows Forms application by starting VB .NET and then choosing the Windows Form template. This displays the default form like the one seen in Figure 23.2.

Next, change the properties of the form as follows:

Size: 500,500
Text: System Information

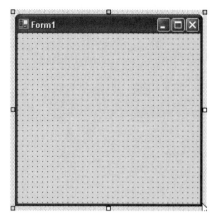

FIGURE 23.2 The default form in VB .NET.

You should see these changes instantly in the IDE (see Figure 23.3).

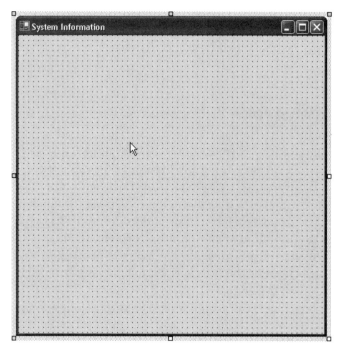

FIGURE 23.3 The form is changed instantly in the IDE.

Next, add a ListBox control to the form. This control will be utilized to store all of the information about the PC, including BIOS, Network, and System. You can add it to the form and then make these changes to it:

Size: 450,450

Location: 10,10

Your form should now look like Figure 23.4.

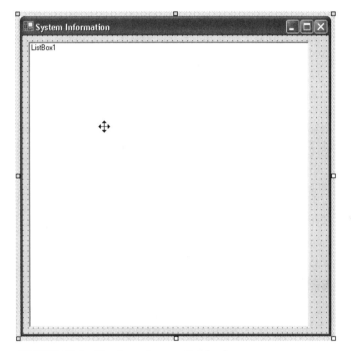

FIGURE 23.4   The form is now finished.

You'll notice that we left the Form1 and ListBox1 names intact. If we had been creating an application that contained more items in its user interface, we probably would have changed these names to reflect their purpose. However, because this only contained a single form and a single control, it wasn't really necessary for our purposes.

WRITING SOME CODE

The next step is to begin writing some code for the application. You can double-click the form to display the Code Editor. Next, add a reference to System.Management and an Imports statement, which is the first line of code in the application and should appear before any other code. The line imports System.Management into the application. Here is the code:

```
Imports System.Management
```

You can add the Imports statement to the Code Editor, which should contain something similar to the following code after you have added it:

```
Imports System.Management

Public Class Form1
:   Inherits System.Windows.Forms.Form

:   Private Sub Form1_Load(ByVal sender As System.Object, ByVal e As
System.EventArgs) Handles MyBase.Load

:   End Sub

End Class
```

WMI REFERENCES

Before we move on, let's take a look at some of the things we can accomplish with WMI and VB .NET. The following list of class names gives you an idea of the various classes that are exposed:

- Win32_X Management classes
- Win32_ComputerSystem
- Win32_DiskDrive
- Win32_LogicalDisk
- Win32_NetworkAdapter
- Win32_NetworkAdapterConfiguration
- Win32_NetworkLoginProfile
- Win32_OperatingSystem
- Win32_Printer

- Win32_Process
- Win32_Processor
- Win32_Service
- Win32_VideoController

A thorough list of the classes follows:

Win32_ComputerSystem class: Allows access to information about Windows PCs; has several domain roles:

0: Standalone workstation

1: Member workstation

2: Standalone server

3: Member server

4: Backup domain controller

5: Primary domain controller

- **Properties**
  Caption: Name of the Windows PC

  PrimaryOwnerName: Win2000 only owner of the PC

  Domain: Name of the domain the PC is a part of

  DomainRole: Role PC has in the domain, see above

  Manufacturer: Manufacturer of the PC

  Model: Model of the PC

  NumberofProcessors: Number of processors in the PC

  SystemType: Processor class of the PC

  SystemStartupDelay: Amount of time in seconds for user to choose OS on startup

  TotalPhysicalMemory: Amount of installed RAM in the system

- **Methods**
  Put_(): Saves changes made to the class

Win32_DiskDrive class: Allows access to disk drive information

- **Properties**
  Caption: Description of the drive

  Description: Brand name of the drive

  InterfaceType: Type of the drive

Manufacturer: Manufacturer of the drive

Partitions: Number of partitions on the drive

Sectors: Sectors of the drive

ScsiBus: SCSI bus number

ScsiTargetID: SCSI ID

Size: Size of the drive in bytes

Win32_LogicalDisk class: Allows access to logical drives

- **Properties**

  DriveType: Type of drive (1: removable, 2: floppy, 3: HD, 4: network, 5: CD-ROM)

  FileSystem: File system of the drive

  Freespace: Amount of free space on disk drive in bytes

  Name: Disk drive letter

  Size: Size of the drive in bytes

  VolumeName: Volume name of the logical disk drive

  VolumeSerialNumber: Serial number of the disk drive

Win32_NetworkAdapter class: Allows access to Windows network adapter properties

- **Properties**

  Description: Name of the network adapter

  MACAddress: MAC address of the network adapter

Win32_NetworkAdapterConfiguration class: Allows access to Windows network adapters (NIC) configuration

- **Collections**

  DefaultIPGateway: Collection of all Default IP gateways

  DNSServerSearchOrder: Collection of all DNS servers IPs

  IPaddress: Collection of all IP addresses for the adapter

  IPsubnet: Collection of all subnet masks

  WinsPrimaryServer: Collection of all WINS servers IPs

- **Properties**

  Description: Name of the network adapter

DHCPenabled: (-1: if DHCP is not enabled, 0: not enabled, 1: DHCP enabled)

DHCPLeaseObtained: Number representing date and time DHCP lease was obtained

DHCPLeaseExpires: Number representing date and time DHCP lease expires

DHCPServer: IP address of the DHCP server

DNSHostname: Name of the host

DNSDomain: Name of the DNS domain

IPenabled: Is true if the adapter has an enabled IPaddress

MACAddress: MAC address of the network adapter

Win32_NetworkLoginProfile class: Allows access to network login profile information

■ **Properties**
LastLogin: Last login of the user
Name: Username of the user

Win32_OperatingSystem class: Allows access to Windows functions

■ **Properties**
BootDevice: Drive that boots the OS

BuildNumber: Build verison of the OS

BuildType: Build type

Caption: Name of the operating system

CSName: Name of the system

CsdVersion: Service pack version

CurrentTimeZone: Time zone

FreePhysicalMemory: Amount of free memory in RAM in KB

FreeVirtualMemory: Amount of free virtual memory in KB

InstallDate: Date the OS was installed

LastBootUpTime: Number representing date and time since PC was last booted up

NumberofProcesses: Number of processes currently running

Organization: Organization that was set when installed

OsLanguage: Number representation of the language of the operating system

Primary: Is true if OS is in use

RegisteredUser: User of the OS

SerialNumber: Serial number of the operating system

SystemDevice: Drive that has the OS files

SystemDirectory: Path to the system directory

TotalVirtualMemorySize: Total size of virtual memory in KB

Version: Version of the operating system

WindowsDirectory: Path to the Windows directory

- **Methods**
  Reboot(): Reboots the PC
  Shutdown(): Shuts down the PC

Win32_Printer class: Allows access to Windows printers

- **Properties**
  Description: Name of the printer

Win32_Process class: Allows access to Windows processes

- **Properties**
  Caption: Name of the Windows process
  CreationDate: String representing date and time process was started
  Name: Name of the Windows process
  Priority: Priority level of the process
  ProcessID: ID of the Windows process
  ThreadCount: Number of process threads
  WorkingSetSize: Amount of memory dedicated to the process in Kb
- **Methods**
  Create(strProcess): Creates a new process
  Terminate(): Terminates the process

Win32_Processor class: Allows information about the processor

- **CPU interfaces**
  1 Other
  2 Unknown

3 Daughter board

4 ZIF socket

5 Replacement

6 None

7 LIF socket

8 Slot1

9 Slot2

10 370 Pin

11 SlotA

12 SlotM

■ **Properties**
AddressWidth: Processor data width in bits

CurrentClockSpeed: Clock speed of the processor

Extclock: External clock speed of the processor

DeviceID: CPU ID of the processor

Description: CPU class, family, model, and stepping of the processor

L2CacheSize: Size of the L2cache on the processor in Kb

L2CacheSpeed: Speed of the L2 cache on the processor

Name: Brand name of the processor

UpgradeMethod: CPU interface, see above

Win32_Service class: Allows access to Windows services

■ **Properties**
Description: Name of the service

DisplayName: Same as description

ServiceType: Type of service, "ShareProcess" or "Own Process"

State: Current state of the service, "Running" or "Stopped"

Status: Current status of the service, "OK"

StartMode: Current Start mode of the service, "Auto," "Manual," or "Disabled"

StartName: Start name of the service, "LocalSystem"

■ **Methods**
StopService(): Stops the service

ChangeStartMode(strMode): Changes the Start mode of the service

Mode: "Automatic" or "Manual"

Win32_VideoController class: Allows access to video properties

■ **Properties**
   Caption: Name of the video card
   CurrentHorizontalResolution: Horizontal resolution of the screen
   CurrentVerticalResolution: Vertical resolution of the screen
   CurrentNumberOfColors: Color depth of the screen

USING THE CLASSES

Now that you have an idea of the classes that we can utilize, we're going to put them to use in this application. Specifically, we're going to use `NetworkAdapterConfiguration`, `BIOS`, and `ComputerSystem`.

Let's begin with a simple property of the ListBox control. We are going to be placing a great deal of information in it, so we'll need to have scroll bars. Let's use the `ScrollAlwaysVisible` property:

```
ListBox1.ScrollAlwaysVisible = True
```

You can add this line to the `Form_Load` event, which should now look like the following code:

```
Private Sub Form1_Load(ByVal sender As System.Object, ByVal e As
System.EventArgs) Handles MyBase.Load

        ListBox1.ScrollAlwaysVisible = True

End Sub
```

The next step is to create some variables for the `ManagementClass` and `ManagementObject`. These are both part of `System.Management` namespace that we imported earlier. Here is the code:

```
Dim mc As Management Class
Dim mo As ManagementObject
```

Next, we assign mc equal to a new `ManagementClass`. The `ManagementClass` in this case is the Network Adapter Configuration (`"Win32_NetworkAdapterConfiguration"`). We can also set moc equal to a `ManagementObjectCollection`. Here is the code for those lines:

```
mc = New ManagementClass("Win32_NetworkAdapterConfiguration")
Dim moc As ManagementObjectCollection = mc.GetInstances()
```

Now, it's time to begin sending information to the listbox. We begin with a couple of lines that will be used to send some basic text information to the listbox. We use its Add method as follows:

```
ListBox1.Items.Add("Network Information")
ListBox1.Items.Add("------------------------------------------------")
```

This information will be used so that the user can quickly look through the list-box to see the information they are looking for. The next step is to use a For Each loop, which is used to repeat a group of statements for each element in an array or, for our particular needs, a collection. We then check to see if IPEnabled is True. If so, we then output the MAC Address and Description properties of the network adapter.

Here is the code:

```
For Each mo In moc
    If mo.Item("IPEnabled") = True Then
        ListBox1.Items.Add("MAC address : " &
mo.Item("MacAddress").ToString())
        ListBox1.Items.Add("Description : " &
mo.Item("Description").ToString())
    End If
Next
```

At this point, we have the network card information, so we add a couple of blank lines to the listbox to help separate the areas. Here is the code for those lines:

```
ListBox1.Items.Add("")
ListBox1.Items.Add("")
```

Now, we turn our attention to the BIOS. This information is obtained through the "Win32_BIOS" management class. We set mc and moc similarly to the earlier step:

```
mc = New ManagementClass("Win32_BIOS")
moc = mc.GetInstances()
```

Again, we use a For Each loop to repeat the code necessary to add the information to the listbox. This time, we use a With statement so that we don't have to type ListBox1.Items for every line in this code. Instead, we can use a With statement to shorten this up. Here is the code:

```
For Each mo In moc
    With ListBox1.Items
    .Add("BIOS Information")
    .Add("-------------------------------------------")
    .Add("Name : " & mo.Item("Name").ToString())
    .Add("Serial Number : " & mo.Item("SerialNumber").ToString())
    .Add("Manufacturer : " & mo.Item("Manufacturer").ToString())
    .Add("Status : " & mo.Item("Status").ToString())
    .Add("Release Date : " & mo.Item("ReleaseDate").ToString())
    .Add("SMBIOS Version : " & mo.Item("SMBIOSBIOSVersion").ToString())
    .Add("SMBIOS Major Version : " &
mo.Item("SMBIOSMajorVersion").ToString())
    .Add("SMBIOS Minor Version : " &
mo.Item("SMBIOSMinorVersion").ToString())
    .Add("SMBIOS Present : " & mo.Item("SMBIOSPresent").ToString())
    .Add("Software Element ID : " &
mo.Item("SoftwareElementID").ToString())
    .Add("Software Element State : " &
mo.Item("SoftwareElementState").ToString())
    .Add("Version : " & mo.Item("Version").ToString())
    .Add("Current Lang. : " & mo.Item("CurrentLanguage").ToString())
    .Add("")
    .Add("")
    End With
Next
```

We now repeat the same process with the ComputerSystem, Processor, and Pointing Device management classes. Here is the code:

```
mc = New ManagementClass("Win32_ComputerSystem")
moc = mc.GetInstances()

For Each mo In moc

With ListBox1.Items
    .Add("Computer System")
    .Add("-------------------------------------------")
    .Add("Caption : " & mo.Item("Caption").ToString())
    .Add("Primary Owner Name : " &
mo.Item("PrimaryOwnerName").ToString())
    .Add("Domain : " & mo.Item("Domain").ToString())
    .Add("Domain Role : " & mo.Item("DomainRole").ToString())
    .Add("Manufacturer : " & mo.Item("Manufacturer").ToString())
    .Add("Model : " & mo.Item("Model").ToString())
    .Add("Number Processors : " &
mo.Item("NumberofProcessors").ToString())
    .Add("System Types : " & mo.Item("SystemType").ToString())
```

```
    .Add("System Startup Delay : " &
mo.Item("SystemStartupDelay").ToString())
    .Add("Physical Memory : " &
mo.Item("TotalPhysicalMemory").ToString())
End With

Next
'Processor Info
        mc = New ManagementClass("Win32_Processor")
        moc = mc.GetInstances()
        For Each mo In moc

            With ListBox1.Items
                .Add("Processor")
                .Add("------------------------------------------------")
                .Add("CPU Name: " & mo.Item("Name").ToString())
                .Add("CPU Voltage Caps : " &
mo.Item("VoltageCaps").ToString())
                .Add("L2 Cache : " & mo.Item("L2CacheSize").ToString())
                .Add("Current Clock Speed : " &
mo.Item("CurrentClockSpeed").ToString())
                .Add("CPU Status : " & mo.Item("CpuStatus").ToString())
                .Add("")
                .Add("")
            End With
        Next

        'Pointing Device
        mc = New ManagementClass("Win32_PointingDevice")
        moc = mc.GetInstances()
        For Each mo In moc

            With ListBox1.Items
                .Add("Pointing Device")
                .Add("-----------------------------------------------")
                .Add("Device ID: " & mo.Item("DeviceID").ToString())
                .Add("Type : " & mo.Item("PointingType").ToString())
                .Add("Manufacturer: " &
mo.Item("Manufacturer").ToString())
                .Add("Number of Buttons: " &
mo.Item("NumberOfButtons").ToString())
                .Add("Status: " & mo.Item("Status").ToString())
                .Add("Caption : " & mo.Item("Caption").ToString())
                .Add("")
                .Add("")

            End With
        Next
```

TESTING THE APPLICATION

At this time, you can test the application by pressing F5, choosing Start from the Debug menu, or clicking the Start button in the IDE. Regardless of your choice, your application opens and should look similar to Figure 23.5.

FIGURE 23.5   The application with information displayed about the PC.

FINAL CODE LISTING

This is the final code listing for the application:

```
Imports System.Management


Public Class Form1
    Inherits System.Windows.Forms.Form

    Private Sub Form1_Load(ByVal sender As System.Object, ByVal e As
System.EventArgs) Handles MyBase.Load
```

```
        ListBox1.ScrollAlwaysVisible = True

        'System.Management
        Dim mc As ManagementClass

        Dim mo As ManagementObject

        'Network Info
        mc = New ManagementClass("Win32_NetworkAdapterConfiguration")
        Dim moc As ManagementObjectCollection = mc.GetInstances()

        ListBox1.Items.Add("Network Information")
        ListBox1.Items.Add("-------------------------------------------
-----")

     For Each mo In moc
          If mo.Item("IPEnabled") = True Then
               ListBox1.Items.Add("MAC address : " &
mo.Item("MacAddress").ToString())
               ListBox1.Items.Add("Description : " &
mo.Item("Description").ToString())
          End If
     Next

        ListBox1.Items.Add("")
        ListBox1.Items.Add("")

        'BIOS Info
        mc = New ManagementClass("Win32_BIOS")
        moc = mc.GetInstances()

        For Each mo In moc

          With ListBox1.Items
               .Add("BIOS Information")
               .Add("--------------------------------------------")
               .Add("Name : " & mo.Item("Name").ToString())
               .Add("Serial Number : " &
mo.Item("SerialNumber").ToString())
               .Add("Manufacturer : " &
mo.Item("Manufacturer").ToString())
               .Add("Status : " & mo.Item("Status").ToString())
               .Add("Release Date : " &
mo.Item("ReleaseDate").ToString())
               .Add("SMBIOS Version : " &
mo.Item("SMBIOSBIOSVersion").ToString())
               .Add("SMBIOS Major Version : " &
mo.Item("SMBIOSMajorVersion").ToString())
```

```
                    .Add("SMBIOS Minor Version : " &
mo.Item("SMBIOSMinorVersion").ToString())
                    .Add("SMBIOS Present : " &
mo.Item("SMBIOSPresent").ToString())
                    .Add("Software Element ID : " &
mo.Item("SoftwareElementID").ToString())
                    .Add("Software Element State : " &
mo.Item("SoftwareElementState").ToString())
                    .Add("Version : " & mo.Item("Version").ToString())
                    .Add("Current Lang. : " &
mo.Item("CurrentLanguage").ToString())
                    .Add("")
                    .Add("")
                End With
            Next


        'Computer Info
        mc = New ManagementClass("Win32_ComputerSystem")
        moc = mc.GetInstances()

        For Each mo In moc

            With ListBox1.Items
                .Add("Computer System")
                .Add("-------------------------------------------------
")
                .Add("Caption : " & mo.Item("Caption").ToString())
                .Add("Primary Owner Name : " &
mo.Item("PrimaryOwnerName").ToString())
                .Add("Domain : " & mo.Item("Domain").ToString())
                .Add("Domain Role : " &
mo.Item("DomainRole").ToString())
                .Add("Manufacturer : " &
mo.Item("Manufacturer").ToString())
                .Add("Model : " & mo.Item("Model").ToString())
                .Add("Number Processors : " &
mo.Item("NumberofProcessors").ToString())
                .Add("System Types : " &
mo.Item("SystemType").ToString())
                .Add("System Startup Delay : " &
mo.Item("SystemStartupDelay").ToString())
                .Add("Physical Memory : " &
mo.Item("TotalPhysicalMemory").ToString())
            End With
        Next
```

```
        End Sub
    End Class
```

SUMMARY

In this chapter, we built a very useful application that can be customized for your particular needs. It would be excellent to include with another project for an About Box or a Troubleshooting Guide. There were several new things we touched on in this example, including a ListBox control, the `With` statement, a `For Each` loop, and obviously the WMI. In Chapter 24, Power Management for the Tablet PC, we build a directory browser using the TreeView control.

24 Power Management for the Tablet PC

As you know, the Tablet PC is often used without an AC power source. In fact, the ability to last for many hours without a charge is one of the strong points for the Tablet PC. It's ironic that a strong point can also be a weak one. With the potential for problems, we need to do what we can to help with power management for the end user. Therefore, we take some time in this chapter to look at the power management API and how we can support some of its many features.

The source code for the projects are located on the CD-ROM in the PROJECTS folder. You can either type them in as you go or you can copy the projects from the CD-ROM to your hard drive for editing.

POWER MANAGEMENT API

Although Windows provides a great deal of power management functions that we can take advantage of, there is currently no way to do it using managed code in VB .NET. Therefore, we'll create a module that will contain the appropriate structures, functions, and API calls.

Let's take a quick look at some of the API functions we'll use:

SetActivePwrScheme: Sets active power scheme

CanUserWritePwrScheme: Determines whether the user can write to power scheme

GetActivePwrScheme: Finds the index of the active power scheme

GetSystemPowerStatus: Locates the power status of the system

ReadGlobalPwrPolicy: Gets the current global power policy settings

ReadProcessorPwrScheme: Gets the processor power policy settings for the specified power scheme

ReadPwrScheme: Gets the power policy settings that are unique to the specified power scheme

WriteProcessorPwrScheme: Writes processor power policy settings for the specified power scheme

We'll also use some of the structures:

GlobalMachinePowerPolicy: This structure contains global computer power policy settings that apply to all power schemes for all users. This structure is part of the next structure called GlobalPowerPolicy.

GlobalPowerPolicy: This structure contains global power policy settings that apply to all power schemes.

GlobalUserPowerPolicy: This is another structure that is part of the GlobalPowerPolicy structure and contains global user power policy settings that apply to all power schemes for a user.

MachinePowerPolicy: This structure, part of the PowerPolicy structure, contains computer power policy settings that are unique to each power scheme on the computer.

MachineProcessorPowerPolicy: This structure contains processor power policy settings that apply while the system is running on AC power or battery power.

PowerActionPolicy: This structure contains information used to set the system power state.

PowerPolicy: This structure contains power policy settings that are unique to each power scheme.

ProcessorPowerPolicy: This structure contains information about processor performance control and C-states.

ProcessorPowerPolicyInfo: This structure is part of ProcessorPowerPolicy and contains information about processor C-state policy settings.

SystemPowerLevel: This structure is part of GlobalUserPowerPolicy and contains information about system battery drain policy settings.

SystemPowerStatus: This structure contains information about the power status of the system.

UserPowerPolicy: This last structure is also part of PowerPolicy. It has power policy settings that are unique to each power scheme for a user.

We need to add these structures and functions to a project. Start a new Windows Forms application and add a new code module to the project called Powerapi.vb. We can now add the code for the structures and functions we have talked about beginning with the structures:

```
Imports System.Runtime.InteropServices
Module powerapi
<StructLayout(LayoutKind.Sequential)> Private Structure
GlobalMachinePowerPolicy
    Public Revision As Integer
    Public LidOpenWakeAc As Integer
    Public LidOpenWakeDc As Integer
    Public BroadcastCapacityResolution As Integer
End Structure

<StructLayout(LayoutKind.Sequential)> Private Structure
GlobalPowerPolicy
    Public User As GlobalUserPowerPolicy
    Public Machine As GlobalMachinePowerPolicy
End Structure

<StructLayout(LayoutKind.Sequential)> Private Structure
GlobalUserPowerPolicy
    Public Revision As Integer
    Public PowerButtonAc As PowerActionPolicy
    Public PowerButtonDc As PowerActionPolicy
    Public SleepButtonAc As PowerActionPolicy
    Public SleepButtonDc As PowerActionPolicy
    Public LidCloseAc As PowerActionPolicy
    Public LidCloseDc As PowerActionPolicy
    Public DischargePolicy0 As SystemPowerLevel
    Public DischargePolicy1 As SystemPowerLevel
    Public DischargePolicy2 As SystemPowerLevel
    Public DischargePolicy3 As SystemPowerLevel
    Public GlobalFlags As Integer
End Structure

<StructLayout(LayoutKind.Sequential)> Private Structure
MachinePowerPolicy
    Public Revision As Integer
    Public MinSleepAc As Integer
    Public MinSleepDc As Integer
    Public ReducedLatencySleepAc As Integer
    Public ReducedLatencySleepDc As Integer
    Public DozeTimeoutAc As Integer
    Public DozeTimeoutDc As Integer
    Public DozeS4TimeoutAc As Integer
    Public DozeS4TimeoutDc As Integer
    Public MinThrottleAc As Byte
    Public MinThrottleDc As Byte
```

```
        Public Pad0 As Byte
        Public Pad1 As Byte
        Public OverThrottledAc As PowerActionPolicy
        Public OverThrottledDc As PowerActionPolicy
    End Structure

    <StructLayout(LayoutKind.Sequential)> Private Structure
    MachineProcessorPowerPolicy
        Public Revision As Integer
        Public ProcessorPolicyAc As ProcessorPowerPolicy
        Public ProcessorPolicyDc As ProcessorPowerPolicy
    End Structure

    <StructLayout(LayoutKind.Sequential)> Private Structure
    PowerActionPolicy
        Public PowerAction As Integer
        Public Flags As Integer
        Public EventCode As Integer
    End Structure

    <StructLayout(LayoutKind.Sequential)> Private Structure PowerPolicy
        Public User As UserPowerPolicy
        Public Machine As MachinePowerPolicy
    End Structure

    <StructLayout(LayoutKind.Sequential)> Private Structure
    ProcessorPowerPolicy
        Public Revision As Integer
        Public DynamicThrottle As Byte
        Public Spare0 As Byte
        Public Spare1 As Byte
        Public Spare2 As Byte
        Public Reserved As Integer
        Public PolicyCount As Integer
        Public Policy0 As ProcessorPowerPolicyInfo
        Public Policy1 As ProcessorPowerPolicyInfo
        Public Policy2 As ProcessorPowerPolicyInfo
    End Structure

    <StructLayout(LayoutKind.Sequential)> Private Structure
    ProcessorPowerPolicyInfo
        Public TimeCheck As Integer
        Public DemoteLimit As Integer
        Public PromoteLimit As Integer
        Public DemotePercent As Byte
        Public PromotePercent As Byte
        Public Spare0 As Byte
```

```
    Public Spare1 As Byte
    Public AllowBits As Integer
End Structure

<StructLayout(LayoutKind.Sequential)> Private Structure
SystemPowerLevel
    Public Enable As Byte
    Public Spare0 As Byte
    Public Spare1 As Byte
    Public Spare2 As Byte
    Public BatteryLevel As Integer
    Public PowerPolicy As PowerActionPolicy
    Public MinSystemState As Integer
End Structure

<StructLayout(LayoutKind.Sequential)> Private Structure
SystemPowerStatus
    Public ACLineStatus As Byte
    Public BatteryFlags As Byte
    Public BatteryLifePercent As Byte
    Public Reserved1 As Byte
    Public BatteryLifeTime As Integer
    Public BatteryFullLifeTime As Integer
End Structure

<StructLayout(LayoutKind.Sequential)> Private Structure UserPowerPolicy
    Public Revision As Integer
    Public IdleAc As PowerActionPolicy
    Public IdleDc As PowerActionPolicy
    Public IdleTimeoutAc As Integer
    Public IdleTimeoutDc As Integer
    Public IdleSensitivityAc As Byte
    Public IdleSensitivityDc As Byte
    Public ThrottlePolicyAc As Byte
    Public ThrottlePolicyDc As Byte
    Public MaxSleepAc As Integer
    Public MaxSleepDc As Integer
    Public Reserved0 As Integer
    Public Reserved1 As Integer
    Public VideoTimeoutAc As Integer
    Public VideoTimeoutDc As Integer
    Public SpindownTimeoutAc As Integer
    Public SpindownTimeoutDc As Integer
    Public OptimizeForPowerAc As Byte
    Public OptimizeForPowerDc As Byte
    Public FanThrottleToleranceAc As Byte
    Public FanThrottleToleranceDc As Byte
```

```
        Public ForcedThrottleAc As Byte
        Public ForcedThrottleDc As Byte
End Structure



Private Class Win32
    Declare Auto Function ApplyPwrScheme Lib "powrprof.dll" Alias
"SetActivePwrScheme" _
        (ByVal SchemeId As Integer, Optional ByVal Unused As Integer =
0, _
        Optional ByVal Unused2 As Integer = 0) As Byte

    Declare Auto Function CanUserWritePwrScheme Lib "powrprof.dll" ()
As Byte

    Declare Auto Function GetActivePwrScheme Lib "powrprof.dll" _
        (ByRef SchemeId As Integer) As Byte

    Declare Auto Function GetSystemPowerStatus Lib "kernel32.dll" _
        (ByRef Status As SystemPowerStatus) As Byte

    Declare Auto Function ReadGlobalPwrPolicy Lib "powrprof.dll" _
        (ByRef GlobalPolicy As GlobalPowerPolicy) As Byte

    Declare Auto Function ReadProcessorPwrScheme Lib "powrprof.dll" _
        (ByVal SchemeId As Integer, ByRef Policy As
MachineProcessorPowerPolicy) As Byte

    Declare Auto Function ReadPwrPolicy Lib "powrprof.dll" Alias
"ReadPwrScheme" _
        (ByVal SchemeId As Integer, ByRef Policy As PowerPolicy) As
Byte

    Declare Auto Function SetActivePwrScheme Lib "powrprof.dll" _
        (ByVal SchemeId As Integer, ByRef GlobalPolicy As
GlobalPowerPolicy, _
        ByRef Policy As PowerPolicy) As Byte

    Declare Auto Function WriteProcessorPwrScheme Lib "powrprof.dll" _
        (ByVal SchemeId As Integer, ByRef Policy As
MachineProcessorPowerPolicy) As Byte

End Class
```

ON THE CD

As you add the code, you probably noticed that we created a class for the API functions. This is an easy way to wrap up the entire set of functions that we'll use throughout the project.

With the API functions added, we can now focus on the functions and Sub procedures we need to create to take advantage of them. The following sections list the functions we will create and their overall objectives along with the code for each of them (the code can be added to the module after the Win32 Class).

GetCpuDynamicThrottling

The first function we look at determines both the AC and DC CPU dynamic throttling values. Here is its code:

```
Public Enum Powermode
    modedc
    modeac
End Enum

Public Enum ThrottlingMode
    ThrottleNone = 0 'NONE
    ThrottleConstant = 1 'CONSTANT
    ThrottleDegrade = 2 'DEGRADE
    ThrottleAdaptive = 3 'ADAPTIVE
End Enum


Private Sub GetCpuDynamicThrottling(ByRef throttleAc As Integer, _
    ByRef throttleDc As Integer)

    Dim schemeId As Integer
    Dim currentPolicy As MachineProcessorPowerPolicy
    Dim result As Byte

    result = Win32.GetActivePwrScheme(schemeId)
    If result <> 1 Then
        Throw New System.Exception("Unable to determine the active
power scheme")
    End If

    result = Win32.ReadProcessorPwrScheme(schemeId, currentPolicy)
    If result <> 1 Then
        Throw New System.Exception("Unable to read the current CPU
power scheme")
```

```
        End If

        throttleAc = currentPolicy.ProcessorPolicyAc.DynamicThrottle
        throttleDc = currentPolicy.ProcessorPolicyDc.DynamicThrottle

    End Sub
```

GetCpuThrottlingAc

The AC throttling power policy is handled in this function. Here is this function's code:

```
Public Function GetCpuThrottlingAc() As ThrottlingMode

    Dim throttleAc As Integer
    Dim throttleDc As Integer

    GetCpuDynamicThrottling(throttleAc, throttleDc)

    Return throttleAc

End Function
```

GetCpuThrottlingDc

This is similar to the previous function, but this one handles the DC throttling power policy. Here is the code:

```
Public Function GetCpuThrottlingDc() As ThrottlingMode

    Dim throttleAc As Integer
    Dim throttleDc As Integer

    GetCpuDynamicThrottling(throttleAc, throttleDc)

    Return throttleDc

End Function
```

GetPowerMode

We need a way to tell if the system is using AC or DC power, which is what we handle with the following code:

```
Public Function GetPowerMode() As Powermode

        Dim pwrStatus As SystemPowerStatus
```

```
        Dim result As Byte

        result = Win32.GetSystemPowerStatus(pwrStatus)
        If result <> 1 Then Throw New System.Exception("Cannot
determine system power status")

        If pwrStatus.ACLineStatus = 0 Then Return Powermode.modedc
        Return Powermode.modeac

    End Function
```

SetCpuDynamicThrottling

This Sub procedure allows us to change the CPU throttling policies for both the AC and DC power modes:

```
    Private Sub SetCpuDynamicThrottling(ByVal throttleAc As Integer, ByVal
    throttleDc As Integer)

        Dim schemeId As Integer
        Dim currentPolicy As MachineProcessorPowerPolicy
        Dim newPolicy As MachineProcessorPowerPolicy
        Dim result As Byte

        result = Win32.CanUserWritePwrScheme()
        If result <> 1 Then
            Throw New System.Exception("Cannot get CPU power information")
        End If

        result = Win32.GetActivePwrScheme(schemeId)
        If result <> 1 Then
            Throw New System.Exception("Cannot determine the active power
scheme")
        End If

        result = Win32.ReadProcessorPwrScheme(schemeId, currentPolicy)
        If result <> 1 Then
            Throw New System.Exception("Cannot read the current CPU power
scheme")
        End If

        newPolicy = currentPolicy
        If throttleAc >= 0 Then newPolicy.ProcessorPolicyAc.DynamicThrottle
= throttleAc
        If throttleDc >= 0 Then newPolicy.ProcessorPolicyDc.DynamicThrottle
= throttleDc

        result = Win32.WriteProcessorPwrScheme(schemeId, newPolicy)
```

```
If result <> 1 Then
    Throw New System.Exception("Cannot change the current CPU power
scheme")
End If

result = Win32.ApplyPwrScheme(schemeId)
If result <> 1 Then
    Win32.WriteProcessorPwrScheme(schemeId, currentPolicy)
    Throw New System.Exception("Cannot apply the changes made to
the power scheme")
End If

End Sub
```

SetCpuThrottling

This Sub procedure changes the CPU throttling power for the current power mode:

```
Public Sub SetCpuThrottling(ByVal throttle As ThrottlingMode)

    Dim pwrMode As Powermode

    pwrMode = GetPowerMode()
    If pwrMode = Powermode.modeac Then
        SetCpuThrottlingAc(throttle)
    Else
        SetCpuThrottlingDc(throttle)
    End If

End Sub
```

SetCpuThrottlingAc

The AC power policy can be changed with this Sub procedure:

```
Public Sub SetCpuThrottlingAc(ByVal throttle As ThrottlingMode)
    SetCpuDynamicThrottling(throttle, -1)
End Sub
```

SetCpuThrottlingDc

This is the same as the previous Sub, but for DC power:

```
Public Sub SetCpuThrottlingDc(ByVal throttle As ThrottlingMode)
    SetCpuDynamicThrottling(-1, throttle)
End Sub
```

That's all there is for the Powerapi.vb module. You can close this and open the form, so that we can create a user interface for our application.

USER INTERFACE

The user interface for this application is going to be quick to create. We need to add the controls shown in Table 24.1 to the form:

TABLE 24.1   Creating the user interface

Type	Name	Text
Label	lblInfo	Info
Label	lblCPU	CPU
StatusBar	StatusBar1	StatusBar1
GroupBox	GroupBox1	CPU

You can refer to Figure 24.1 for their locations.

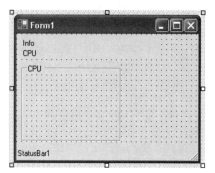

FIGURE 24.1   The controls in the correct location.

The next step is to add the following radio buttons (shown in Table 24.2), placing them inside GroupBox1, which automatically allows the end user the ability to select only one of the options.

TABLE 24.2 Adding radio buttons to the user interface

Name	Text
NoCPUThrottling	No Throttling
ConstThrottling	Constant Throttling
DegradeLow	Low Power Degrade
Adaptive	Adaptive

You can refer to Figure 24.2 for the visible GUI. The last control we need to add to the project is a Timer. You can leave its settings as the default, with the exception of the Interval property, which needs to be set to 10000.

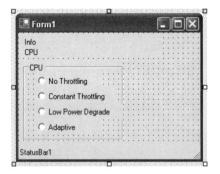

FIGURE 24.2 The visible GUI is finished.

We can now create the code that will actually be used in our application. It calls the functions we created earlier. Open the Code Editor for Form1. Begin with the Imports statements and the following declarations:

```
Imports Microsoft.Win32
Imports System.Management
Dim moReturn As ManagementObjectCollection
Dim moSearch As ManagementObjectSearcher
Dim mo As ManagementObject
```

We are going to display the CPU speed in the status bar, which is the reason for the declarations and the Timer control. Let's look at the Timer Elapsed method.

This `Sub` procedure will be executed every time the timer gets to 10 seconds. During execution, we'll display the processor information.

Here is the code:

```
Private Sub Timer1_Elapsed(ByVal sender As System.Object, ByVal e As
System.Timers.ElapsedEventArgs) Handles Timer1.Elapsed
    moSearch = New Management.ManagementObjectSearcher("Select * from
Win32_Processor")
    moReturn = moSearch.Get

    For Each mo In moReturn
        Dim strOut As String
        Dim speed As UInt32
        speed = mo("CurrentClockSpeed")
        strOut = mo("Name") + " - " + speed.ToString
        StatusBar1.Text = mo("Name") + " - " + speed.ToString & " mhz"
    Next
End Sub
```

The `Form_Load` event is next. We use this `Sub` procedure to hook the `SystemEvents PowerModeChange` event handler function. We need to do this so that it is called when the power mode is changed. We'll also call it immediately so as to initialize the current AC or DC throttling information.

Here is the code for the procedure:

```
Private Sub Form1_Load(ByVal sender As System.Object, ByVal e As
System.EventArgs) Handles MyBase.Load


    Dim cpuMode As ThrottlingMode

    AddHandler Microsoft.Win32.SystemEvents.PowerModeChanged, AddressOf
Me.OnPowerModeChange
    OnPowerModeChange(Me, New
PowerModeChangedEventArgs(PowerModes.StatusChange))
    StatusBar1.Text = "Retrieving CPU Info"

    Select Case cpuMode

        Case ThrottlingMode.ThrottleNone
            NoCpuThrottle.Select()

        Case ThrottlingMode.ThrottleConstant
            ConstThrottling.Select()

        Case ThrottlingMode.ThrottleDegrade
```

```
            DegradeLow.Select()

        Case ThrottlingMode.ThrottleAdaptive
            Adaptive.Select()

    End Select
End Sub
```

As we just used it, let's look at `OnPowerModeChange`, which handles the system power mode change event. The `PowerModes.StatusChange` event is the only event we need to concern ourselves with. We begin by checking this event to see if it is a `StatusChange`. If so, we continue on in the `Sub` procedure. Otherwise, we continue our execution and ignore the event.

Here is the code:

```
Private Sub OnPowerModeChange(ByVal sender As Object, ByVal args As
PowerModeChangedEventArgs)
 Dim pwrMode As Powermode
 Dim cpuMode As ThrottlingMode
 If args.Mode <> PowerModes.StatusChange Then Return
```

The remaining steps read the current power mode, and depending on what it is, we read the CPU throttling mode. The following code finishes the procedure:

```
Try

        pwrMode = GetPowerMode()

        If pwrMode = Powermode.modeac Then
            cpuMode = GetCpuThrottlingAc()
        Else
            cpuMode - GetCpuThrottlingDc()
        End If

    Catch except As System.Exception

        MessageBox.Show(except.Message.ToString(), "Error ",
MessageBoxButtons.OK, MessageBoxIcon.Error)

        Return

    End Try


        If pwrMode = Powermode.modeac Then
```

```
        lblInfo.Text = "Using AC power"
    Else
        lblInfo.Text = "Using DC power"
    End If

    Select Case cpuMode

        Case ThrottlingMode.ThrottleNone
            lblCPU.Text = "CPU is unthrottled"

        Case ThrottlingMode.ThrottleConstant
            lblCPU.Text = "CPU is constantly throttled"

        Case ThrottlingMode.ThrottleDegrade
            lblCPU.Text = "CPU throttles on low battery"

        Case ThrottlingMode.ThrottleAdaptive
            lblCPU.Text = "CPU is throttled adaptively"

    End Select
```

The final coding steps for this project involve the radio buttons. These buttons simply call the powerapi to set the appropriate type of throttling depending on the button clicked.

Here is the code:

```
Private Sub ConstThrottling_CheckedChanged(ByVal sender As
System.Object, ByVal e As System.EventArgs) Handles
ConstThrottling.CheckedChanged
    powerapi.SetCpuThrottling(powerapi.ThrottlingMode.ThrottleConstant)
    OnPowerModeChange(Me, New
PowerModeChangedEventArgs(PowerModes.StatusChange))
End Sub


Private Sub Adaptive_CheckedChanged(ByVal sender As System.Object,
ByVal e As System.EventArgs) Handles Adaptive.CheckedChanged
    powerapi.SetCpuThrottling(powerapi.ThrottlingMode.ThrottleAdaptive)
    OnPowerModeChange(Me, New
PowerModeChangedEventArgs(PowerModes.StatusChange))
End Sub


Private Sub NoCpuThrottle_CheckedChanged(ByVal sender As System.Object,
ByVal e As System.EventArgs) Handles NoCpuThrottle.CheckedChanged
    powerapi.SetCpuThrottling(powerapi.ThrottlingMode.ThrottleNone)
```

```
      OnPowerModeChange(Me, New
PowerModeChangedEventArgs(PowerModes.StatusChange))
End Sub


Private Sub DegradeLow_CheckedChanged(ByVal sender As System.Object,
ByVal e As System.EventArgs) Handles DegradeLow.CheckedChanged
      powerapi.SetCpuThrottling(powerapi.ThrottlingMode.ThrottleDegrade)
      OnPowerModeChange(Me, New
PowerModeChangedEventArgs(PowerModes.StatusChange))
End Sub
```

TESTING THE APPLICATION

You can now save the application and test it. Figure 24.3 displays the application as an example of how it should appear. You can test the various options to see how well it performs.

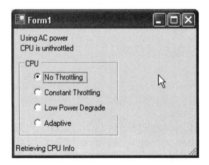

FIGURE 24.3   The application being executed.

SUMMARY

This chapter continued our look at the hardware of the Tablet PC. We created a project that you could include in your applications to help an end user manage their power management, one of the more important aspects of mobile computing. In Chapter 25, Virtual Joystick, we build a "virtual joystick" as an example of how games or applications can be controlled via the pen.

25 Virtual Joystick

In this chapter, we take a look at the basics of building a utility for simulating a joystick. The basic concepts behind the joystick are to eliminate the need for a real joystick to be connected to the Tablet PC to play games. In theory, this sounds great, but realistically, the usefulness of this application is probably very limited depending on the game. With that said, you can use this same technique for controlling your own game projects (we'll do something similar in Chapter 26, Pong Game), and with some tweaking, this could be an interesting way to input information into any application without typing.

ON THE CD

The source code for the projects are located on the CD-ROM in the PROJECTS folder. You can either type them in as you go or you can copy the projects from the CD-ROM to your hard drive for editing.

KEY CONCEPTS

As you already know, the general idea behind this application is to simulate a joystick. Most games that support joysticks also offer the user the ability to use a keyboard. Obviously, there might be times that a Tablet PC user may not have his keyboard with him, but might want to play a game. The obvious approach to playing the game is to use the Input Panel, but this obstructs so much of the screen's real estate that it might be difficult to see the game you are playing.

Our approach is to develop an application based around the SendKeys method. By using SendKeys, we have the ability to send keystrokes to other applications. We can assign keys for the various directions and buttons that a joystick would normally use (see Figure 25.1). The keys correspond to the keys the game is set up to use. We'll store the keys in the Registry and retrieve them as needed.

FIGURE 25.1 The joystick will look like this.

GETTING STARTED

ON THE CD

Begin by creating a new VB Window Forms application. As you can see in Figure 25.1, the GUI is quite simple. We use a background image on a form to simulate the appearance of a joystick. The image is available on the CD-ROM that is included with the book in the Chapter 25 project folder. After assigning the background to the image, resize the form to approximate its dimensions. The next step is to place label controls to represent the keys.

The following Label controls need to be created with the following names and placed in the correct locations:

- lblUp
- lbl45
- lblRight
- lbl135
- lblDown
- lbl225
- lblLeft
- lbl315

The previous labels handle the directional elements for the joystick. There are a few remaining labels that we need to add. First, we need to add labels called lblReturnPos and lblDelay, which help to control how long the application waits between virtual keystrokes. The last two steps are to add a Timer control, a StatusBar control, and an InkEdit control to the form. The InkEdit control will be used to change the values of the directions as the user will not have access to his keyboard.

You can leave their names as the default. The form should now look like Figure 25.2 (so that you can see them better, the labels have had values assigned to them).

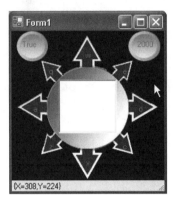

FIGURE 25.2   The final GUI.

WRITING CODE

We begin with importing the appropriate namespaces. Add the following code to the Code Editor:

```
Imports Microsoft.Win32
Imports Microsoft.Ink
Imports System.Windows.Forms.Screen
```

We are now going to create the variables for our application, which include values for the directions, the screen dimensions for both x and y directions, a Boolean value for joystick enabled, and the Registry key value.

Here is the code:

```
Dim DUp, DDown, DLeft, DRight, D45, D135, D225, D315, DReturnPos,
DDelay As Object
    Dim X As Integer = PrimaryScreen.Bounds.Width / 2
    Dim Y As Integer = PrimaryScreen.Bounds.Height / 2
    Dim JEnabled As Boolean

    Dim pRegKey As RegistryKey =
Registry.LocalMachine.OpenSubKey("SOFTWARE\\VirtualJoystick")
```

THE REGISTRY

The Windows Registry is a central database for application configuration settings and other information required by the applications. In fact, this is its only purpose. We are going to manually add some values to the Registry so that we can see how this information is stored. For this application, you could enter values into the Registry manually, or you can simply run the Chapter25.Reg file included on the CD-ROM in the Chapter 25 folder. This adds the values to the Registry automatically for you.

ON THE CD

NOTE

Making changes to the Registry can cause serious problems with your system. Refer to an appropriate text before making changes that you don't already understand. With this in mind, it is recommended that you enter the Registry values with the Chapter25.reg file.

After the values are entered, you can run the Registry Editor by choosing Run from the Start menu. Use the filename regedit, and then click OK to open it (see Figure 25.3). Depending on your previous use of the editor, your Registry may look very different from the one displayed in Figure 25.3. You can see from the figure that the Registry is hierarchical data storage for the settings and has five main keys listed under My Computer. You can look up the "VirtualJoystick" application if you want. Either way, now that you have some idea of the values stored, you can close the Registry Editor.

The .NET Framework provides two classes (Registry and RegistryKey) to work with the Registry. These classes are defined in the Microsoft.Win32 namespace, which is why we added a reference to it earlier.

The Registry Class

The Registry class is the first of the two classes we look at and contains members that provide access to Registry keys. We can define Registry keys in the following order:

CurrentUser: Stores information about user preferences

LocalMachine: Stores configuration information for the local machine

ClassesRoot: Stores information about types (and classes) and their properties

Users: Stores information about the default user configuration

PerformanceData: Stores performance information for software components

CurrentConfig: Stores non-user-specific hardware information

DynData: Stores dynamic data

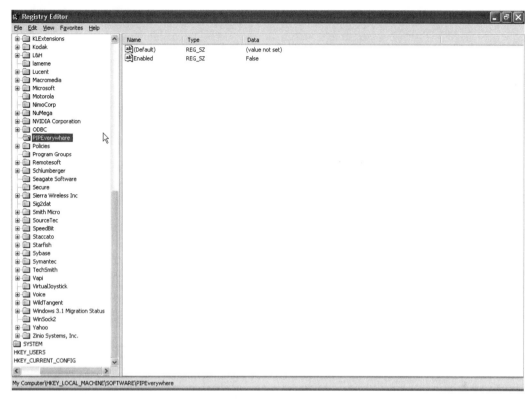

FIGURE 25.3 The Windows Registry Editor.

As you noticed with the code we added, our data will be stored in Local-Machine under "VirtualJoystick." The Registry class has a field corresponding to each of these key types. The Registry class members are described in the following list:

ClassesRoot: Provides access to HKEY_CLASSES_ROOT key

CurrentConfig: Provides access to HKEY_CURRENT_CONFIG key

CurrentUser: Provides access to HKEY_CURRENT_USER key

DynData: Provides access to HKEY_DYN_DATA key

LocalMachine: Provides access to HKEY_LOCAL_MACHINE key

PerformanceData: Provides access to HKEY_PERFORMANCE_DATA key

Users: Provides access to HKEY_USERS key

As you can see from our code, if you want to access the HKEY_LOCAL_ MACHINE key, you need to call the `Registry.LocalMachine` member, which returns a `RegistryKey` type.

The `RegistryKey` Class

The `RegistryKey` class contains members that allow us to add, remove, replace, and read Registry data. Some of its common properties are detailed in the following list:

`Name:` Represents the name of the key

`SubKeyCount:` Represents the count of subkeys at the base level, for current key

`ValueCount:` Represents the count of values in the key

Some of the common methods of the `RegistryKey` class are detailed in the following list:

`Close:` Closes the key

`CreateSubKey:` Creates a new subkey if doesn't exist; otherwise, opens an existing subkey

`DeleteSubKey:` Deletes the specified subkey

`DeleteSubKeyTree:` Deletes a subkey and any children

`DeleteValue:` Deletes the specified value from a key

`GetSubKeyNames:` Returns an array of strings that contains all the subkey names

`GetValue:` Returns the specified value

`GetValueNames:` Retrieves an array of strings that contains all the value names associated with this key

`OpenSubKey:` Opens a subkey

`SetValue:` Sets the specified value

INITIALIZING VALUES

In the `Form_Load` event, we call the `Init` procedure, which we are about to create. The `Init` procedure gets values from the Registry. We use the `GetValue` method, which returns the value of a subkey in the form of Object. We then initialize the values displayed by the labels so that the user can visually see the values. Initializing the values for the labels calls the `InitLabels` procedure.

Here are the two procedures:

```
Private Sub Init()
    DUp = pRegKey.GetValue("Up")
    DDown = pRegKey.GetValue("Down")
    DLeft = pRegKey.GetValue("Left")
    DRight = pRegKey.GetValue("Right")
    D45 = pRegKey.GetValue("45")
    D135 = pRegKey.GetValue("135")
    D225 = pRegKey.GetValue("225")
    D315 = pRegKey.GetValue("315")
    DReturnPos = pRegKey.GetValue("ReturnPos")   ' Setup delay & return
position
    DDelay = pRegKey.GetValue("Delay")
    pRegKey.Close()

    JEnabled = False
    Timer1.Interval = Int(DDelay)

    InitLabels(DUp, lblUp)
    InitLabels(DDown, lblDown)
    InitLabels(DLeft, lblLeft)
    InitLabels(DRight, lblRight)
    InitLabels(D45, lbl45)
    InitLabels(D135, lbl135)
    InitLabels(D225, lbl225)
    InitLabels(D315, lbl315)
    InitLabels(DReturnPos, lblReturnPos)
    InitLabels(DDelay, lblDelay)

    InkEdit1.Text = ""

End Sub

Private Sub InitLabels(ByVal name As Object, ByVal lbl As Label)
    lbl.BackColor = Color.Transparent
    lbl.ForeColor = Color.Red
    lbl.TextAlign = ContentAlignment.MiddleCenter
    lbl.Text = name
End Sub
```

You may have noticed that we set the Timer1 Interval property equal to DDelay. This is the value we retrieved from the Registry. By default, this value is equal to 2000 milliseconds. It's now time to write the code that sends the keys to other applications using the Timer1_Elapsed event.

We begin by testing to see if the application is enabled. If so, we then use a series of `If...Then` and `Case` statements to determine where the mouse is positioned on the screen. Depending on where our pen position is on the screen, we send the appropriate keys. For example, if our pen is positioned in the middle of the screen, our joystick would be centered and would not send a key. If we were to move down, the application would then send the value of down.

Here is the code:

```
Private Sub Timer1_Elapsed(ByVal sender As System.Object, ByVal e As
System.Timers.ElapsedEventArgs) Handles Timer1.Elapsed
    StatusBar1.Text = Control.MousePosition.ToString

    If JEnabled Then

        If Control.MousePosition.X > X + 50 And Control.MousePosition.Y
< Y - 50 Then
            SendKeys.Send(D45)
            Exit Sub
        ElseIf Control.MousePosition.X > X + 50 And
Control.MousePosition.Y > Y + 50 Then
            SendKeys.Send(D135)
            Exit Sub
        ElseIf Control.MousePosition.X < X - 50 And
Control.MousePosition.Y > Y + 50 Then
            SendKeys.Send(D225)
            Exit Sub
        ElseIf Control.MousePosition.X < X - 50 And
Control.MousePosition.Y < Y - 50 Then
            SendKeys.Send(D315)
            Exit Sub
        End If

        Select Case Control.MousePosition.X
            Case Is < X - 100
                SendKeys.Send(DLeft)
                Exit Sub
            Case Is > X + 100
                SendKeys.Send(DRight)
                Exit Sub
        End Select

        Select Case Control.MousePosition.Y
            Case Is < Y - 100
                SendKeys.Send(DUp)
                Exit Sub
            Case Is > Y + 100
```

```
                    SendKeys.Send(DDown)
                    Exit Sub
            End Select

        End If

    End Sub
```

There are a couple of things we have left to do. For starters, we need to have a way to instruct the application when it is enabled. An easy way to do this is to use the Form_MouseDown event and test for the right mouse button being pressed. If the right button is pressed, the application will set JEnabled to the opposite of itself, effectively setting it from True to False or False to True.

Here is the code:

```
Private Sub Form1_MouseDown(ByVal sender As Object, ByVal e As
System.Windows.Forms.MouseEventArgs) Handles MyBase.MouseDown
    If e.Button = MouseButtons.Right Then
        JEnabled = Not (JEnabled)
    End If
End Sub
```

The final step in the application is to look at how we can store new values for the various directions. We can use the SetValue method to set these values, but first, we need a way to change them. Each of the labels has a Click event and we can use the Click event for all of them. For this example, we look at only lblUp. This is also where the InkEdit control comes into play.

We begin this procedure by testing to see if the length of InkEdits text is equal to 1. If so, we assume that the value needs to be stored. Otherwise, we can assume the user entered an incorrect value because we only have the ability to store a single character. We now store the value and then set the InkEdit control to an empty string, finishing by closing the Registry.

Here is the code:

```
    Private Sub lblUp_Click(ByVal sender As System.Object, ByVal e As
System.EventArgs) Handles lblUp.Click
        Dim regKey As RegistryKey
        Dim ver As Decimal

        If InkEdit1.TextLength = 1 Then
regKey = Registry.LocalMachine.OpenSubKey("SOFTWARE\\VirtualJoystick",
True)
            lblUp.Text = InkEdit1.Text
```

```
            InkEdit1.Text = ""
            regKey.SetValue("Up", lblUp.Text)
            regKey.Close()
        End If
    End Sub
```

The application is now finished. You can test it in several ways. An easy way is to open Notepad and then run the application. Next, right-click the application to enable it and then maximize Notepad. Depending on your pen's position, you should see the various keys being input into Notepad, as shown in Figure 25.4.

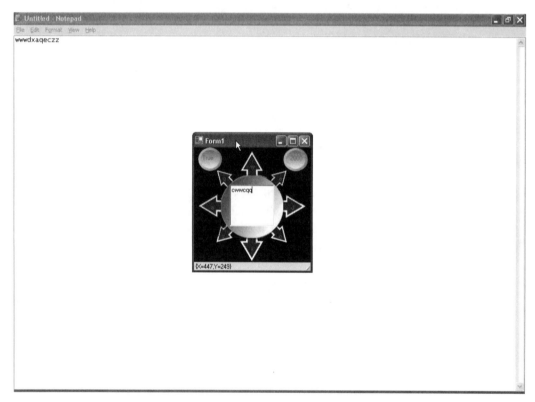

FIGURE 25.4 The application is sending keys to Notepad.

SUMMARY

In this chapter, we built a very unique application that emulates a joystick. We used a variety of new concepts, including reading and writing to the Registry. The application could easily be upgraded by adding code to create a taskbar icon for the application. You could also add the ability to alter the strings for each direction and save them to the Registry. Another interesting idea is to turn this application into a full-screen virtual keyboard, which would send keys to an application depending on the position of the mouse. In Chapter 26, we build a pong game that uses some of the concepts we've looked at for controlling movement.

26 Pong Game

In this chapter, we continue in the general direction we started with in our last application by developing our first game. The game is a simple game of Pong (see Figure 26.1) using very basic drawing techniques and simple Artificial Intelligence (AI).

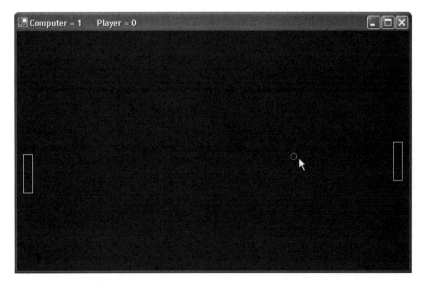

FIGURE 26.1 The game we'll create is Pong.

ON THE CD

The source code for the projects are located on the CD-ROM in the PROJECTS folder. You can either type them in as you go or you can copy the projects from the CD-ROM to your hard drive for editing.

GAME OVERVIEW

All game projects, be it for the Tablet PC or a standard desktop PC, have a few concepts and are constructed in a similar manner. These are a few steps that are consistent among game projects:

1. **Initialization:** Initializes the key objects, such as starting locations of players, number of lives, score, direction, and so forth
2. **Game Loop:** Controls movement, score, and collisions for every frame of a game
3. **Exit:** Exits the game, stores high scores, asks if user wants to play again, and so on

Let's start a new Windows Forms application in VB and add the following variables to the project:

```
Dim ball As Single
Dim x, y, dx, dy As Single
Dim xpaddle1, ypaddle1, dxpaddle1, dypaddle1 As Single
Dim xpaddle2, ypaddle2, dxpaddle2, dypaddle2, paddleWidth, paddleHeight
As Single
Dim blnGameOn As Boolean
Dim t As New Threading.Thread(AddressOf game)
Dim g As Graphics
Dim Computer As Integer
Dim Player As Integer
Dim MoveDown As Boolean
Dim MoveUp As Boolean
```

GAME LOOP

All of the action for our game takes place in a loop, but before we look through the procedure that will house the loop, let's take a minute to get the rest of the application out of the way.

The first step is to use the `Form1_Activated` event to start the threading for our game loop. Threads are a powerful abstraction for allowing parallelized operations. For example, graphical updates can happen while a different thread is performing computations. Conceptually, threads are pretty simple to understand, but they can get very complicated to manage from a programming standpoint. In this example, we'll have a very simple implementation:

```
Private Sub Form1_Activated(ByVal sender As Object, ByVal e As
System.EventArgs) Handles MyBase.Activated
    t.Start()
End Sub
```

The `Form_Load` event comes after the `Form_Activated` event. We use the `Form_Load` event to set the `Text` property to show the scores for the game (at this time 0 for the computer and the player), and we set the variables `Player` and `Computer` to 0.

Here is the procedure:

```
Private Sub Form1_Load(ByVal sender As System.Object, ByVal e As
System.EventArgs) Handles MyBase.Load
    Me.Text = "Computer = " & "0" & "        Player = " & "0"
    Player = 0
    Computer = 0
End Sub
```

We will now implement a simple movement method similar to the virtual joystick we created in the previous chapter. We can test the pen's position and move the paddle depending on its position. Here is the code:

```
Private Sub Form1_MouseMove(ByVal sender As Object, ByVal e As
System.Windows.Forms.MouseEventArgs) Handles MyBase.MouseMove
    If e.Y > (Me.Height / 2) + 50 Then
        MoveDown = True
        MoveUp = False
        Exit Sub
    ElseIf e.Y < (Me.Height / 2) - 50 Then
        MoveUp = True
        MoveDown = False
        Exit Sub
    Else
        MoveUp = False
        MoveDown = False
    End If
End Sub
```

The final event we deal with before moving on to the loop thread is to set up what occurs when the form is closing. We set the Boolean variable `blnGameOn` equal to `False` so that the loop will end.

Here is the code:

```
Private Sub Form1_Closing(ByVal sender As Object, ByVal e As
System.ComponentModel.CancelEventArgs) Handles MyBase.Closing
    blnGameOn = False
End Sub
```

The remaining text in this chapter is used to go through the various parts of the game loop. It is the job of this loop to draw every frame of the game, creating the illusion of movement by repositioning each object after each frame is drawn.

The first part of the game procedure sets the size of the ball, the position of the ball, and the size and position of the computer controlled paddle (paddle1) and the player controlled paddle (paddle2).

Here is the code:

```
Private Sub game()
    Randomize()
    ball = 10 'Size
    x = Rnd() * (Me.Width - ball)
    y = Rnd() * (Me.Height - 30 - ball)
    dx = 1
    dy = 1

    paddleWidth = 15
    paddleHeight = 60
    xpaddle1 = 10
    xpaddle2 = Me.Width - paddleWidth - 20
    ypaddle1 = Me.Height / 2 - paddleHeight / 2
    ypaddle2 = Me.Height / 2 - paddleHeight / 2
    dxpaddle1 = 0
    dxpaddle2 = 0
    dypaddle1 = 1
    dypaddle2 = -1

    g = Me.CreateGraphics

    blnGameOn = True
```

The game loop is a While loop, which continues executing as long as blnGameOn is equal to True. The loop consists of a series of steps, as follows:

1. Calculate the new position of objects.
2. Draw the object at the calculated location.
3. Provide the user with time to see the frame.
4. Delete the frame.
5. Return to Step 1.

The objects in our game, a ball and two paddles, are moved according to four variables:

x: Controls horizontal position of the ball

y: Controls the vertical position of the ball

dx: Controls the direction and magnitude of movement of the object along the horizontal axis

dy: Controls the direction and magnitude of movement of the object along the vertical axis

To move the objects, we use the x, y, dx, and dy values of the objects with the following calculations:

x = x + dx

y = y + dy

After we calculate the new coordinate values, then we draw the object at this new location. There are a few additional things we need to check as we move the ball and paddles. That is, we need to calculate when there is a collision between the ball and either paddle, or the ball and the upper and lower bounds of the form, which results in reversing the direction of the ball. The last collision we need to watch is the collision between the ball and either end. This adds a point to the appropriate player and positions the items where they began. To reverse the direction of the object along its dx and dy, we just multiply these values by -1. Therefore if dy = 1 and the ball is moving downwards, then we can use dy = dy * -1 to change dy's value to -1, which in turn, makes the object move up.

There are many different algorithms involved in determining when two objects collide on a computer screen; we will look at one of the simpler, albeit inaccurate, ones. We calculate the distance between the centers of two objects on the screen. If it is less than the sum of the radius of the two objects, then the objects are going to collide.

If you remember back to the steps for our game loop, we have covered the first step. Now, it's time to draw the objects. For the ball, it's as simple as using the following:

```
g.DrawEllipse(New Pen(Color.Orange), x, y, ball, ball
```

Similarly, we can draw the paddles:

```
g.DrawRectangle(New Pen(Color.White), xpaddle1, ypaddle1, paddleWidth,
paddleHeight)
g.DrawRectangle(New Pen(Color.Yellow), xpaddle2, ypaddle2, paddleWidth,
paddleHeight)
```

We now pause the application so the user can see how it appears on screen. We use the thread's Sleep method to delay the application. Next, we clear the graphics and then update the Text property of the form to reflect any new changes to the score.

Here is the complete procedure:

```
While (blnGameOn)
    x = x + dx
    y = y + dy
    If ((x + ball) >= Me.Width) Then
        dx = dx * -1
        Computer = Computer + 1
        x = Rnd() * (Me.Width - ball)
        y = Rnd() * (Me.Height - 30 - ball)
        dx = 1
        dy = 1
    End If

    If (x <= 0) Then
        dx = dx * -1
        Player = Player + 1
        x = Rnd() * (Me.Width - ball)
        y = Rnd() * (Me.Height - 30 - ball)
        dx = 1
        dy = 1
    End If

    If ((y + ball) >= (Me.Height - 30) Or y <= 0) Then
        dy = dy * -1
    End If

    g.DrawEllipse(New Pen(Color.Orange), x, y, ball, ball)

    xpaddle1 = xpaddle1 + dxpaddle1
    xpaddle2 = xpaddle2 + dxpaddle2
    ypaddle1 = ypaddle1 + dypaddle1
    ypaddle2 = ypaddle2 + dypaddle2

    If (ypaddle1 <= y Or (ypaddle1 + paddleHeight) >= (Me.Height -
30)) Then
        dypaddle1 = 1
    End If
```

```
        If (ypaddle1 >= y Or (ypaddle1 + paddleHeight) >= (Me.Height -
30)) Then
            dypaddle1 = -1
        End If

        If ypaddle2 >= Me.Height - 35 - paddleHeight Then
            ypaddle2 = Me.Height - 35 - paddleHeight
        ElseIf ypaddle2 <= 0 Then
            ypaddle2 = 0
        End If

        If MoveDown Then
            dypaddle2 = 1
        ElseIf MoveUp Then
            dypaddle2 = -1
        Else
            dypaddle2 = 0
        End If


        g.DrawRectangle(New Pen(Color.White), xpaddle1, ypaddle1,
paddleWidth, paddleHeight)
        g.DrawRectangle(New Pen(Color.Yellow), xpaddle2, ypaddle2,
paddleWidth, paddleHeight)

        If Math.Abs((x + ball / 2) - (xpaddle1 + paddleWidth / 2)) <
(ball / 2 + paddleWidth / 2) And Math.Abs((y + ball / 2) - (ypaddle1 +
paddleHeight / 2)) < (ball / 2 + paddleHeight / 2) Then
            dx = dx * -1
            x = xpaddle1 + paddleWidth + 1
        ElseIf Math.Abs((x + ball / 2) - (xpaddle2 + paddleWidth / 2))
< (ball / 2 + paddleWidth / 2) And Math.Abs((y + ball / 2) - (ypaddle2
+ paddleHeight / 2)) < (ball / 2 + paddleHeight / 2) Then
            dx = dx * -1
            x = xpaddle2 - paddleWidth - ball - 1
        End If

        t.Sleep(10)

        g.Clear(Color.Black)

        Me.Text = "Computer = " & Computer.ToString & "      Player = "
& Player.ToString

    End While

End Sub
```

You can now test the application, which should look like Figure 26.2 when it is executed.

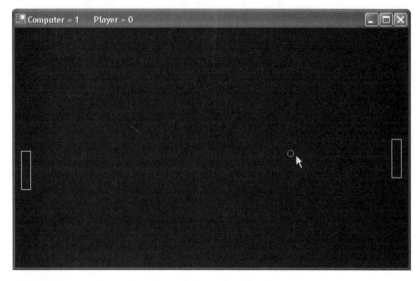

FIGURE 26.2   The application is being executed.

SUMMARY

In this chapter, we built our first game for the Tablet PC. Although simple, it is a good introduction to some of the techniques you'll use with many of the games you can create. This game also used a technique similar to the virtual joystick for the movement of the player-controlled paddle. In Chapter 27, Not Quite a Magic Ball, we build another entertainment application that simulates a magic ball.

27 Not Quite a Magic Ball

I n this chapter, we continue looking at the entertainment aspects of Tablet PC development as we create a "Magic Ball," which answers any questions we give it. You are probably familiar with the general ideas of the application, which gives generic answers to any question we ask it.

ON THE CD

The source code for the projects are located on the CD-ROM in the PROJECTS folder. You can either type them in as you go or you can copy the projects from the CD-ROM to your hard drive for editing.

STARTING THE PROJECT

Let's start a new VB Windows Forms application and begin this project by constructing the GUI, which consists of a couple of PictureBox controls, a ListBox control, and a Button control. You can see their properties in the Table 27.1 and their respective locations in Figure 27.1.

TABLE 27.1 Adding controls to the application

Type	Name	Location	Size
PictureBox	picDest	8,8	400,400
PictureBox	PictureBox1	n/a	n/a
ListBox	ListBox1	n/a	n/a
Button	Button1	425,375	80,25
Timer	tmrSpin		

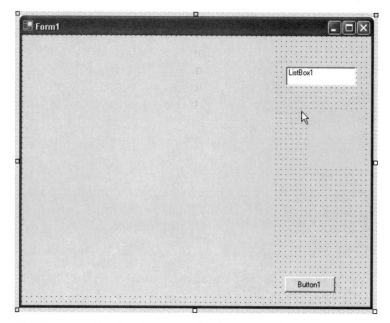

FIGURE 27.1 The GUI for our application is complete.

With the GUI in place, we can concentrate on the programming aspects of the project. Begin by adding the following variable declarations:

```
Private SourceBm As Bitmap
Private SourceWid As Integer
Private SourceHgt As Integer
Private SourceCx As Single
Private SourceCy As Single
Private SourceCorners As PointF()

Private DestWid As Integer
Private DestHgt As Integer
Private DestCx As Single
Private DestCy As Single
Private DestBm As Bitmap
Private DestBackColor As Color

Dim Positive As Boolean

Dim txtAnswer As String

Dim PicTop As Integer
```

The first section of these variables handles the source image we are going to load. This is an image of the ball, which is located in the Projects folder on the CD-ROM in the Chapter 27 directory. This image is loaded into PictureBox1 as a way of getting it into the application. This PictureBox is never visible, nor will the end user ever be aware of its existence.

We can load the image into the PictureBox in the Form_Load event. The code for this part of the event, which loads the image and sets some of its properties, has been used several times, so it's not worth repeating at this late stage in the book:

```
Me.BackColor = Color.White
        picDest.SizeMode = PictureBoxSizeMode.CenterImage
        PictureBox1.Image =
Image.FromFile(System.Windows.Forms.Application.StartupPath &
"\ball.bmp")
        SourceBm = New Bitmap(PictureBox1.Image)
        SourceWid = SourceBm.Width
        SourceHgt = SourceBm.Height
        SourceCx = SourceWid / 2
        SourceCy = SourceHgt / 2
        SourceCorners = New PointF() { _
            New PointF(0, 0), _
            New PointF(SourceWid, 0), _
            New PointF(0, SourceHgt), _
            New PointF(SourceWid, SourceHgt)}
```

The only thing that would be different to you in the previous code deals with the corners of the image. We use the points to rotate the image, which appears to give it some basic animation. The next step is to load the image into the destination variables, which are the second set of the original variables. This is the variable that will be manipulated and ultimately loaded into the picDest picture box. We also need to translate the corners to center the box at origin:

```
DestWid = Math.Sqrt(SourceWid * SourceWid + _
    SourceHgt * SourceHgt)
DestHgt = DestWid
DestCx = DestWid / 2
DestCy = DestHgt / 2
DestBm = New Bitmap(DestWid, DestHgt)

Dim i As Long
For i = 0 To 3
    SourceCorners(i).X -= SourceCx
    SourceCorners(i).Y -= SourceCy
Next i
```

The remaining portion of the procedure sets some additional properties, such as setting the Interval of the timer to 100 and its Enabled property to False, initializing stop_time to Now, setting the text of Button1 to "Spin", and setting the text of the form to "Not Too Magic Ball". Additionally, we add items to the ListBox, which are the answers the ball gives us when we stop it from spinning.

Here is the remaining code for the procedure:

```
DestBackColor = picDest.BackColor

tmrSpin.Interval = 100

tmrSpin.Enabled = False
Button1.Text = "Spin"

ListBox1.Items.Add("No")
ListBox1.Items.Add("Could be")
ListBox1.Items.Add("Possible")
ListBox1.Items.Add("Don't Know")
ListBox1.Items.Add("Try Again")
ListBox1.Visible = False
PictureBox1.Visible = False
Me.Text = "Not Too Magic Ball"
PicTop = picDest.Top
```

The button is used to start and stop the ball from spinning. It is within its Click event that we set this up using an If...Then statement. When the user clicks the button, the If...Then checks to see if tmrSpin is enabled. If it is, then we know that the ball is spinning and we need to stop it, change the Text property of the button to "Spin", and produce an answer. If it is not enabled, we can start the spinning, change the text of the button to "Stop", and set the answer to an empty string.

Here is the entire procedure:

```
Private Sub Button1_Click(ByVal sender As System.Object, ByVal e As
System.EventArgs) Handles Button1.Click
    Dim answer As Integer

    If tmrSpin.Enabled Then
        tmrSpin.Enabled = False
        Button1.Text = "Spin"
        picDest.Image = PictureBox1.Image.Clone
        answer = CInt(Rnd(1) * 4)
        txtAnswer = ListBox1.Items(answer)

    Else
```

```
        tmrSpin.Enabled = True
        Button1.Text = "Stop"
        txtAnswer = ""
    End If

End Sub
```

There are two remaining procedures we need to work with, the first of which is the Paint event of picDest. We can use this Paint event to provide the ability to draw text on a picture box programmatically. This is how the answer will be drawn, and why we have to set it to an empty string whenever we are spinning the ball.

Here is the code for the Paint event:

```
Private Sub picDest_Paint(ByVal sender As Object, ByVal e As
System.Windows.Forms.PaintEventArgs) Handles picDest.Paint
    Dim g As Graphics = e.Graphics
    g.DrawString(txtAnswer, Me.Font, New SolidBrush(Color.Black), 170,
185)
End Sub
```

The final procedure we need to work with is the tmrSpin elapsed procedure. Within this procedure, we are going to perform the calculations for rotating and moving the picture so that it appears to be somewhat animated. The first lines of code begin by testing to see if we have a positive or negative value stored in the Boolean variable called Positive. This variable is used to go from positive to negative, negative to positive, and so on and so forth. By going back and forth with random values, it appears as if the ball is being twisted back and forth.

Here is the code for the beginning part of the tmrSpin elapsed procedure:

```
Dim dtheta = (Rnd(10) * 10) * Math.PI / 180.0

If Positive Then
    dtheta = -dtheta
    Positive = False
Else
    Positive = True
End If

Static theta As Single = 0
theta += dtheta
Dim corners() As PointF = SourceCorners

ReDim Preserve corners(2)
```

To rotate the image, we use the built-in `Sin` and `Cos` functions, translate to center the results of the destination image, and then display the results using `picDest`. `picDest` is also moved, which simply adds to the illusion of movement, by changing its `Top` property. When the button is clicked to stop the procedure, `picDest` is returned to its starting point and the original image is loaded back into it so that the ball appears exactly as it started. This also makes rendering the answer much easier.

Here is the code:

```
Dim sin_theta As Single = Math.Sin(theta)
Dim cos_theta As Single = Math.Cos(theta)
Dim X As Single
Dim Y As Single
Dim i As Long
For i = 0 To 2
    X = corners(i).X
    Y = corners(i).Y
    corners(i).X = X * cos_theta + Y * sin_theta
    corners(i).Y = -X * sin_theta + Y * cos_theta
Next i

For i = 0 To 2
    corners(i).X += DestCx
    corners(i).Y += DestCy
Next i

Dim gr_out As Graphics = Graphics.FromImage(DestBm)
gr_out.Clear(DestBackColor)
gr_out.DrawImage(SourceBm, corners)

If Positive Then
    picDest.Top = picDest.Top - Rnd(10)
Else
    picDest.Top = PicTop
End If

picDest.Image = DestBm
```

This actually finishes the application, which can now be saved and tested. When you run it, make sure the ball rotates back and forth as in Figures 27.2 and 27.3. You can also verify that you receive an answer similar to the one seen in Figure 27.4.

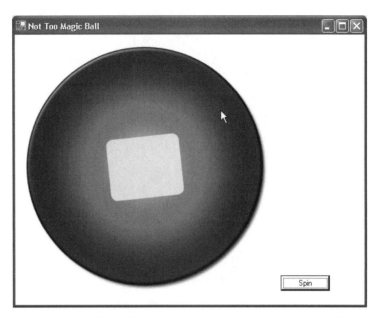

FIGURE 27.2 The ball rotates one direction to start an animation.

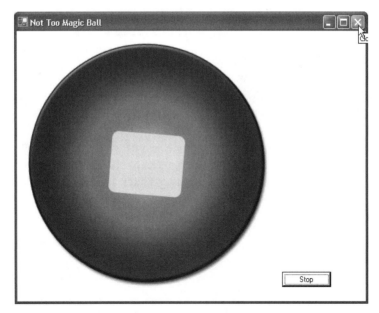

FIGURE 27.3 The ball moves the opposite direction to simulate back and forth movement.

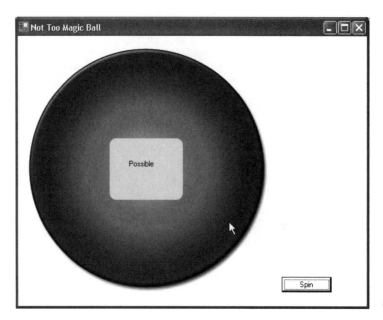

FIGURE 27.4 The animation is stopped and the answer is displayed.

SUMMARY

In this chapter, we created another entertainment style of application. Although the Tablet PC is such a new platform, it appears that simple types of games are probably best suited for its hardware as mobility and battery power are an issue. That said, users of a Tablet PC might like to play games when they have power connected to their machine or when they know they can recharge their systems before they need to use it again for serious business purposes. Like the PDA market, the Tablet PC is poised for great growth in the area of entertainment and game development. With that in mind, we move our focus back to these topics, including 3D rendering with OpenGL and DirectX, in Chapter 29, 3D Rendering with OpenGL and DirectX 9. In Chapter 28, Storing Ink in a Database, we go back to the world of ink and look at how we can store ink in a database.

28 Storing Ink in a Database

In the previous chapter, we continued our work on entertainment-related soft-ware. In this chapter, we take a break from this and return back to digital ink, and specifically, how to store ink in a database. This is a very important objec-tive because database applications can now store such ink as a signature, making many applications that were once impossible a reality.

ON THE CD

The source code for the projects are located on the CD-ROM in the PROJECTS folder. You can either type them in as you go or you can copy the projects from the CD-ROM to your hard drive for editing.

DATABASE OVERVIEW

If you are familiar with data access in previous versions of VB, then you are prob-ably already familiar with ActiveX Data Objects (ADO) or the Data Control. In VB .NET, the Data Control is no longer available, and ADO.NET has replaced ADO.

Creating the Form

ON THE CD

The first step is to create the database using Microsoft Access. If you don't have Ac-cess, you can use the database included on the CD-ROM in the Chapter 28 Project folder. If you have Access, and want to create the database, you need to name it Worksig.mdb and include the fields shown in Table 28.1 in a table called Main:

Open VB .NET, create a new Windows Forms application, and click the Data tab in the ToolBox window. Drag an OleDbDataAdapter object onto the form, which causes the Data Adapter Configuration Wizard to appear. Click Next on the Wizard's initial frame (see Figure 28.1).

TABLE 28.1   Adding fields to the table called Main

Field	DataType
Id	AutoNumber
Customer	Text
Date	Date/Time
Ink	Memo
Description	Memo
Parts	Memo

FIGURE 28.1   The Data Adapter Configuration Wizard.

In the next window, you choose your data connection. As we don't currently have one, click the New Connection button, which brings you to a Data Link Properties window.

Click the Provider tab and choose Microsoft Jet 4.0 OLE DB Provider, and then click the Next button. At this time, you should be on the Connection tab. In the Dataset name text box, you can browse your system to find the Worksig.mdb database. The other options are fine as is.

Click on the Test Connection button and be sure you get "Test Connection Succeeded." If you have problems, go back through the steps until you find the error. After you are successful with the test connect, click OK on the Data Link Properties window.

Click the Next button on the Data Adapter Configuration Wizard's window. The next window is the Choose a Query Type window. Select "Use SQL Statements" from the window and then click Next. On the Generate SQL Statements window, click the Query Builder button. Add the Main table and then close the Tables window. Click on all the fields and then click OK. View the SQL statements that were generated for us. Click Next and see that the `Select` statement and table mappings are OK, and commands are generated. Click the Finish button to end the wizard.

You should now find yourself back in the VB application window. You can rename the `OleDbDataAdpapter` object to `odaSample`. Right-click on `odaSample`, and from the context menu, choose Generate Dataset. Leave the default settings in the Generate Dataset dialog box, but change the name from DataSet1 to pdsSample, and click OK. You now have a persistent file or data set (this is the reason we gave the file a prefix of "pds") called pdsSample.xsd, which is now visible in the Properties window. This file contains an XML schema that describes your data set and allows you to treat your dynamic data set as a file at design time. It's amazingly simple and effective.

You also have a new object visible below the form with the other data objects called pdsSample1. You can also change its name to dsSample, which stands for data set Sample. Making the names easier to remember also makes it much easier when binding controls or when you reference a project that is several months old.

Binding Controls

We now have access to the database, but we still need a way to add and retrieve information from it. We construct our GUI out of standard controls and then bind the controls to the appropriate fields. Add four TextBox controls to the form called txtCustomer, txtWork, txtParts, and txtSignature. Figure 28.2 shows the fields and the general layout of the form. The txtWork and txtParts controls are much larger than the other two and need to have their `Multiline` properties set to `True`. You can clear the `Text` properties for all of them. Click on the txtCustomer text box, and in the Properties window, drill down on its `DataBindings` property. Next, choose Text, and click the drop-down list arrow. Drill down through the dsSample node, then on the Main node, and choose Customer from the list by double-clicking it. The same operation can be performed for the remaining TextBox controls, making sure to bind them to the appropriate field in the database.

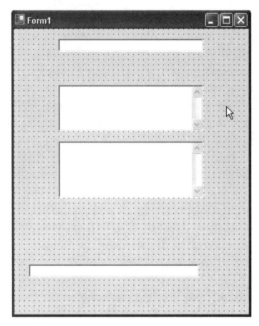

FIGURE 28.2 The basic layout of the form
with four TextBox controls added.

It's easy to tell that the TextBox controls have all been bound to a field as you will see a small gold-colored cylinder representing the data-binding icon next to their Text properties in the Properties window.

The next control to add is a DatePick control. You can place it beneath txt-Customer and above txtWork. This control is also bound to a field, this time the Date field. The process is identical to the TextBox controls, so you can do this now.

The DatePick control is the final control that we bind to a field in the database, although we do need to add the controls shown in Table 28.2 at this time:

TABLE 28.2 Adding additional controls

Type	Name	Text
GroupBox	gbSignature	Signature
Button	btnPrevious	Previous

TABLE 28.2 Adding additional controls (*continued*)

Type	Name	Text
Button	btnSave	Save
Button	btnAddNew	Add New
Button	btnNext	Next
Label	lblNumber	0
Button	btnClear	X

You can view their placement in Figure 28.3, which also displays additional Label controls that are used only for the end users' benefit. That is, they do not play any programmatic role in the example. You should also add the labels using the figure as an example.

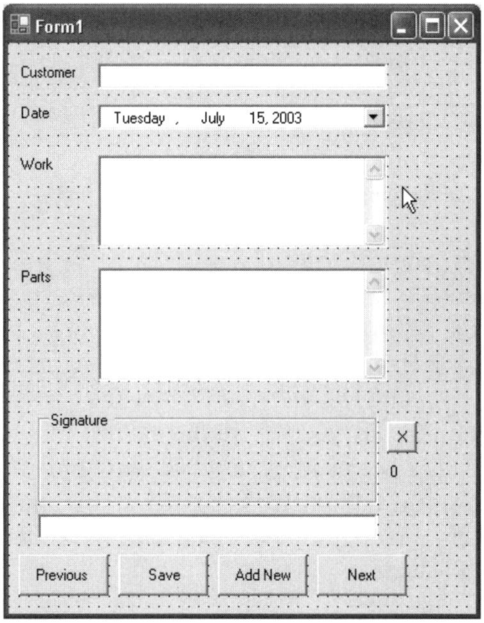

FIGURE 28.3 The final GUI.

You may have noticed that we have bound the txtSignature control to a field in the database but have not bound the gbSignature GroupBox, which is the area in which ink is displayed and written. This is because it's much easier to save the string data representation of the ink and then render it as needed. We look at this in more detail when we get to the step later in the chapter.

There is only one line of code to write to fill the data controls with the sample data from the database. Open the Code Editor, and using the `Form_Load` event, add the following two lines of code, one of which simply sets the form's text:

```
odaSample.Fill(DsSample, "main")
Me.Text = "Work Order Database"
```

You can now save the application, run it, and the first record (if any) from the data set should be visible. However, we cannot navigate the database or really do much of anything with the data at this time. Let's begin to add the navigation code.

NAVIGATING THE DATABASE

The code to navigate the database is as simple as the lines of code we have written. We begin with btnNext. In the `btnNext_Click` event, we add the code for navigating the database. If we were not attempting to deal with rendering the ink, this would require only a single line of code:

```
Me.BindingContext(DsSample, "main").Position += 1
```

However, this is obviously not the case. We begin the procedure by refreshing gbSignature. Next, we check the `lblNumber.Text` property to see if it is less than the total available records in the database -1 (we are looking at the previous record so that we know if we have reached 0). If it is, we then move the position in the database up a record and assign lblNumber to the current position in the database. If it is not less than the last record, we can assume that we have reached the end of the data set and display a message box that says "End of Data." The final step is to test the contents of the txtSignature text box. If it is an empty string, we can ignore ink for the current record. Otherwise, we need to render the ink.

Here is the code:

```
Private Sub btnNext_Click(ByVal sender As System.Object, ByVal e As
System.EventArgs) Handles btnNext.Click
    gbSignature.Refresh()
    If lblNumber.Text < Me.BindingContext(DsSample, "main").Count - 1
Then
```

```
        Me.BindingContext(DsSample, "main").Position += 1
        lblNumber.Text = Me.BindingContext(DsSample, "main").Position
    Else
        MsgBox("End of Data", MsgBoxStyle.OKOnly, "Ink Work Order")
    End If
    If txtSignature.Text <> "" Then RenderInk()
End Sub
```

Rendering the ink is the most interesting aspect of this particular application, but before we look at it in further detail, we need to add a reference to the Ink SDK and the following `Imports` statements to the application:

```
Imports Microsoft.Ink
Imports System
Imports System.Drawing
Imports System.Windows.Forms
```

It's also a good time to add the global variables, which include the following:

```
Public myInk() As Byte
Public myInkString As String
Dim WithEvents myInk2 As InkCollector
Dim thePenInputPanel As New PenInputPanel()
```

You probably can remember back to the earlier chapters where we have already dealt with the PenInputPanel and the InkCollector. We are not going to spend much time on either of these. The `InkCollector_Load` procedure is listed next:

```
Private Sub InkCollection_Load(ByVal sender As System.Object, ByVal e
As System.EventArgs) Handles MyBase.Load
    myInk2 = New InkCollector(gbSignature.Handle)
    myInk2.Handle = gbSignature.Handle
    myInk2.Enabled = True
End Sub
```

After all of that, we can finally get back to the `RenderInk` procedure. It simply takes the text from txtSignature, stores it in myInkString, and then creates the ink from the Base64String we store in the database.

Here is the code:

```
Private Sub RenderInk()
    Dim ink As New Microsoft.Ink.Ink()
    Dim r As New Microsoft.Ink.Renderer()
    Dim g As Graphics
```

```
        myInkString = txtSignature.Text
        g = gbSignature.CreateGraphics()
        ink.Load(Convert.FromBase64String(myInkString))
        r.Draw(g, ink.Strokes)
    End Sub
```

Now that we have the RenderInk Sub procedure finished, we can go back and add the ink rendering to the Form_Load event. Without this, if the first database record has ink, it is not visible on the screen.

Here is the now complete code:

```
Private Sub Form1_Load(ByVal sender As System.Object, ByVal e As
System.EventArgs) Handles MyBase.Load
    odaSample.Fill(DsSample, "main")
    Me.Text = "Work Order Database"
    lblNumber.Text = Me.BindingContext(DsSample, "main").Position
    If txtSignature.Text <> "" Then RenderInk()
End Sub
```

The remaining code for the application is similar to what we have already done. The following procedures assign the Pen Input Panel to all of the text boxes:

```
Private Sub btnClear_Click(ByVal sender As System.Object, ByVal e As
System.EventArgs) Handles btnClear.Click
    gbSignature.Refresh()
    txtSignature.Text = ""
End Sub

Private Sub txtCustomer_Enter(ByVal sender As Object, ByVal e As
System.EventArgs) Handles txtCustomer.Enter
    thePenInputPanel.AttachedEditControl = txtCustomer
End Sub

Private Sub txtWork_Enter(ByVal sender As Object, ByVal e As
System.EventArgs) Handles txtWork.Enter
    thePenInputPanel.AttachedEditControl = txtWork
End Sub

Private Sub txtParts_Enter(ByVal sender As Object, ByVal e As
System.EventArgs) Handles txtParts.Enter
    thePenInputPanel.AttachedEditControl = txtParts
End Sub
```

The only remaining code to write is for btnSave, btnAddNew, and btnPrevious. btnPrevious works similarly to btnNext, so we don't really need to spend time on it. btnSave is the most interesting of the procedures. It begins by converting the ink to a base64 string and storing the string in myInkString. It then deletes the strokes from myInk2 and checks the value of myInkString to make sure that it does not equal "AAA=". If it does not equal "AAA=", then we save the string value so that we can later retrieve it and display it as Ink. The last of the procedures is btnAddNew, and with a few lines of code, it should be self-explanatory at this time. Here is the code for each of the procedures:

```
Private Sub btnSave_Click(ByVal sender As System.Object, ByVal e As
System.EventArgs) Handles btnSave.Click
    myInk =
myInk2.Ink.Save(Microsoft.Ink.PersistenceFormat.InkSerializedFormat)
    myInkString = System.Convert.ToBase64String(myInk)
    myInk2.Ink.DeleteStrokes()

    If myInkString.ToString <> "AAA=" Then
        txtSignature.Text = myInkString.ToString
    End If

    Me.BindingContext(DsSample, "Main").EndCurrentEdit()
    odaSample.Update(DsSample, "main")

    MsgBox("Database Updated Successfully", MsgBoxStyle.OKOnly, "Save
Record")
End Sub

Private Sub btnAddNew_Click(ByVal sender As System.Object, ByVal e As
System.EventArgs) Handles btnAddNew.Click
    Me.BindingContext(DsSample, "Main").EndCurrentEdit()
    Me.BindingContext(DsSample, "Main").AddNew()
    lblNumber.Text = Me.BindingContext(DsSample, "main").Position
    txtSignature.Text = ""
    gbSignature.Refresh()
End Sub

Private Sub btnPrevious_Click(ByVal sender As System.Object, ByVal e As
System.EventArgs) Handles btnPrevious.Click
    gbSignature.Refresh()
    If lblNumber.Text > 0 Then
        Me.BindingContext(DsSample, "main").Position -= 1
        lblNumber.Text = Me.BindingContext(DsSample, "main").Position
    Else
        MsgBox("Start of Data", MsgBoxStyle.OKOnly, "Ink Work Order")
        RenderInk()
    End If
    If txtSignature.Text <> "" Then RenderInk()
```

```
End Sub
```

You can now test the application, making sure to check out the navigation and ink rendering. Figure 28.4 displays the application with rendering working.

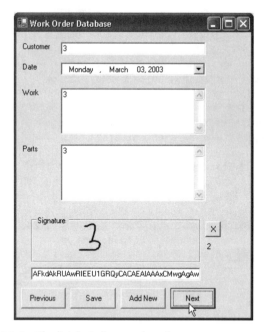

FIGURE 28.4   The ink is being rendered.

SUMMARY

In this chapter, we built an application that stored ink in a database—something that is useful for a variety of applications. Rather than storing the ink directly, the ink is stored with a text representation and then rendered in a GroupBox. In Chapter 29, 3D Rendering with OpenGL and DirectX 9, we look at 3D rendering on the Tablet PC.

29 3D Rendering with OpenGL and DirectX 9

In this chapter, we look at the two major 3D APIs that we have access to on the Tablet PC: OpenGL® and DirectX®. In Chapter 30, Using Third-Party Engines, we explore a third-party engine that wraps up many of the DirectX functions, making the development process a much quicker and easier task.

ON THE CD

The source code for the projects are located on the CD-ROM in the PROJECTS folder. You can either type them in as you go or you can copy the projects from the CD-ROM to your hard drive for editing.

OPENGL WITH THE TAO LIBRARY

Before we develop with OpenGL, we are going to download an open source library that is a wrapper for OpenGL. To get started, download the library, currently at version 0.1.1, from *http://www.randyridge.com/Tao/*. Once downloaded, you need to unzip the compressed file. Inside the extracted directory, you will find two files. These files should be placed in a location that you will remember because we need to add a reference to the DLL from our project.

Because we are using Windows, and Glut is not installed by default, you will now need to download and install it. You can download the Windows binary from *http://www.xmission.com/~nate/glut.html*. Inside the compressed file, you will find Glut32.dll among others. The DLL file needs to be copied to C:\Windows\ System32.

Create a new empty application project in VB .NET and a reference to the Tao DLL that we downloaded earlier. If you are familiar with OpenGL, you have undoubtedly heard about the Redbook (*http://www.opengl.org/developers/code/ examples/redbook/*). If you look at the page, and browse approximately one fourth of the way down, you will see an application called light. We are going to develop an application similar to light so that you can compare the code for each of them. It also allows you to see how quick and easy Tao is to work with.

Add a new empty Class Library to the project, leaving the name as Class1.vb. Next, add the following Imports to the Class Library:

```
Imports Tao.OpenGl
```

The next step is to create Sub Main. This is the first Sub procedure that gets called in the application. We begin this Sub by initializing Glut. Next, we create a window of size 500 x 500, set its position, and give the window a name. The remaining lines of code for this procedure call various additional Sub procedures that we have yet to create.

Here is the code:

```
Public Shared Sub Main(ByVal args() As String)
    Glut.glutInit()
    Glut.glutInitDisplayMode(Glut.GLUT_SINGLE Or Glut.GLUT_RGB Or
Glut.GLUT_DEPTH)
    Glut.glutInitWindowSize(500, 500)
    Glut.glutInitWindowPosition(100, 100)
    Glut.glutCreateWindow("Light")
    Init()
    Glut.glutDisplayFunc(New Glut.DisplayCallback(AddressOf Display))
    Glut.glutKeyboardFunc(New Glut.KeyboardCallback(AddressOf
Keyboard))
    Glut.glutMainLoop()
End Sub
```

The first of the Sub procedures that is called from Sub Main is Init. This procedure sets the material, light position, shading mode, and, generally speaking, a variety of attributes related to the lights and the 3D sphere we'll create.

Here is the code:

```
Private Shared Sub Init()
        Dim materialSpecular() As Single = {1.0F, 1.0F, 1.0F, 1.0F}

    Dim materialShininess() As Single = {50.0F}

    Dim lightPosition() As Single = {1.0F, 1.0F, 1.0F, 0.0F}

    Gl.glClearColor(0.0F, 0.0F, 0.0F, 0.0F)
    Gl.glShadeModel(Gl.GL_SMOOTH)
```

```
Gl.glMaterialfv(Gl.GL_FRONT, Gl.GL_SPECULAR, materialSpecular)
Gl.glMaterialfv(Gl.GL_FRONT, Gl.GL_SHININESS, materialShininess)
Gl.glLightfv(Gl.GL_LIGHT0, Gl.GL_POSITION, lightPosition)

Gl.glEnable(Gl.GL_LIGHTING)
Gl.glEnable(Gl.GL_LIGHT0)
Gl.glEnable(Gl.GL_DEPTH_TEST)
End Sub
```

Because we mentioned it, it's now time to create the sphere in the Sub procedure called Display:

```
Private Shared Sub Display()
    Gl.glClear(Gl.GL_COLOR_BUFFER_BIT Or Gl.GL_DEPTH_BUFFER_BIT)
    Glut.glutSolidSphere(0.5, 20, 16)
    Gl.glFlush()
End Sub
```

The next procedure that is called is Keyboard. It is simply used to test for the press of the Esc key. If the key is pressed, the environment ends. Otherwise, code continues to function as normal.

Here is the code:

```
Private Shared Sub Keyboard(ByVal key As Byte, ByVal x As Integer,
ByVal y As Integer)
    Select Case key
        Case 27
            Environment.Exit(0)
            Exit Sub
    End Select
End Sub
```

That's all there is to this application, which displays a sphere, as shown in Figure 29.1. This is a simple example, but you can quickly see how powerful Tao can be to you as you develop for the Tablet PC. In the next section of this chapter, we look at Direct X 9.

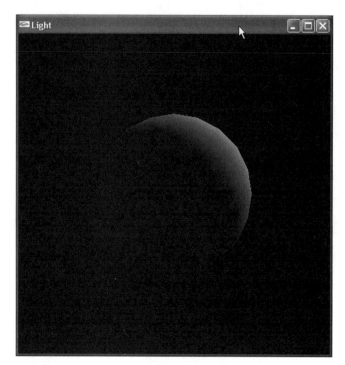

FIGURE 29.1 This sphere is visible on execution.

DIRECTX 9 MANAGED CODE

We have looked at OpenGL, which is available on many platforms and is a very popular 3D API. Now, we take a look at DirectX, which is available only for Windows PCs. As we are working with the Tablet PC in this book, this does not have any issues for us. However, if you ever plan to run an application outside of Windows, DirectX will probably never be available. The idea behind managed code means that if the .NET Framework becomes available for another platform, your application will instantly run on it as long as it is using all managed code. It's true that this helps some applications, but probably will never affect DirectX programs because DirectX is tied to the Windows OS and most likely will never be ported. On the other hand, if you were to develop a Tao-based application, you should have much more portability.

Like OpenGL, we need to get the appropriate SDK before we can begin writing software. For DirectX, you can freely download it from *www.microsoft.com*. It is currently in release 9.0a.

The first step is to create a new Windows Forms application in VB. We'll use the form for rendering our 3D sphere, which looks very similar to the OpenGL version we have already developed. We can leave the properties of the default form alone and concentrate on the code.

First, add references to `Microsoft.DirectX`, `Microsoft.DirectX.Direct3D`, and `Microsoft.Direct3DX`, and add the following `Imports`:

```
Imports Microsoft.DirectX
Imports Microsoft.DirectX.Direct3D
```

The next step is setting up the variables for the application:

```
Private DX9 As Microsoft.DirectX.Direct3D.Device
Private Present As Direct3D.PresentParameters
Private Mesh As Direct3D.Mesh
Private Material As Direct3D.Material
Private Timer1 As Timer
```

In the `Form_Load` event, we check for an available adapter, create a device to render to, and initialize Direct3D. At the end of the procedure, we make calls to two additional procedures, which are completed next. Here is the code for this procedure:

```
    Private Sub Form1_Load(ByVal sender As System.Object, ByVal e As
System.EventArgs) Handles MyBase.Load
        Me.Text = "DirectX Light"
        Dim objAdapters As Direct3D.AdapterInformation

        objAdapters = Direct3D.Manager.Adapters(0)

        Present = New Direct3D.PresentParameters()
        With Present
            .Windowed = True
            .SwapEffect = Direct3D.SwapEffect.Discard
            .BackBufferFormat = objAdapters.CurrentDisplayMode.Format
            .EnableAutoDepthStencil = True
            .AutoDepthStencilFormat = Direct3D.DepthFormat.D16
        End With

        DX9 = New Direct3D.Device(objAdapters.Adapter, _
            Direct3D.DeviceType.Hardware, Me, _
            Direct3D.CreateFlags.HardwareVertexProcessing, Present)

        AddHandler DX9.DeviceReset, AddressOf Me.OnDeviceReset
```

```
            InitializeDirect3D()
            StartRefreshCycle()
    End Sub
```

The `InitializeDirect3D` procedure basically just makes calls to other procedures, including `CreateMesh`, `CreateMaterial`, `CreateLights`, and `InitializeView`. The names of the procedures reflect the actions that they perform. The `StartRefreshCycle` procedure also does what you would expect. Here is the code for the two procedures:

```
Private Sub InitializeDirect3D()
    CreateMesh()
    CreateMaterials()
    CreateLights()
    InitializeView()
End Sub

Private Sub StartRefreshCycle()
    Timer1 = New Timer()
    Timer1.Enabled = True
    Timer1.Interval = 20
    AddHandler Timer1.Tick, AddressOf Me.Render
    Timer1.Start()
End Sub
```

We now create each of the `Sub` procedures called by the `InitializeDirect3D` procedure, beginning with `CreateMesh`. The `CreateMesh` procedure creates a sphere, which is rendered in the scene much like the sphere we created in the OpenGL version. Here is the code:

```
Private Sub CreateMesh()
    Mesh = Direct3D.Mesh.Sphere(DX9, 15, 15, 15)
End Sub
```

The next step is to create the material for the mesh. We can use the `CreateMaterials` `Sub` procedure, which was called next by `InitializeDirect3D`. We are not going to use a texture for the mesh and will simply use the color white for the material. Along the same lines, the code for creating the light is located in the `CreateLights` `Sub` procedure. The code could have been combined, but for easier reading, they have been placed in their own procedures. Here is the code for each of them:

```
Private Sub CreateMaterials()
    Material = New Direct3D.Material()
    Material.Diffuse = Color.White
```

```
End Sub

Private Sub CreateLights()
    Dim Light0 As Direct3D.Light = DX9.Lights(0)
    Light0.Type = Direct3D.LightType.Directional
    Light0.Direction = New Vector3(0, -1, 1)
    Light0.Diffuse = Color.White
    Light0.Ambient = Color.Gray
    Light0.Enabled = True
    Light0.Commit()

    DX9.RenderState.Lighting = True
    DX9.RenderState.Ambient = Color.Gray
End Sub
```

The next procedure that gets called is `InitializeView`. Like the name suggests, the code basically sets up the view for our scene. Here is the code:

```
Private Sub InitializeView()
    Dim eyePosition As New Vector3(0, 0, -75)
    Dim direction As New Vector3(0, 0, 0)
    Dim upDirection As New Vector3(0, 1, 0)

    Dim view As Matrix = Matrix.LookAtLH(eyePosition, direction,
upDirection)
    DX9.SetTransform(Direct3D.TransformType.View, view)

    Dim fieldOfView As Single = Math.PI / 4
    Dim aspectRatio As Single = 1.0
    Dim nearPlane As Single = 1.0
    Dim farPlane As Single = 500.0

    Dim projection As Matrix = _
        Matrix.PerspectiveFovLH(fieldOfView, aspectRatio, nearPlane,
farPlane)

    DX9.SetTransform(Direct3D.TransformType.Projection, projection)
End Sub
```

The remaining two procedures are `Render` and `OnDeviceReset`. `OnDeviceReset` simply calls `InitializeDirect3D`, the procedure we looked at earlier. The `Render` procedure contains the code, which actually causes the mesh to be rendered to the screen. Here is the code for both of them:

```
Private Sub OnDeviceReset(ByVal Sender As Object, ByVal e As EventArgs)
    InitializeDirect3D()
```

```
End Sub

Private Sub Render(ByVal sender As Object, ByVal e As EventArgs)
    DX9.Clear(Direct3D.ClearFlags.Target Or
Direct3D.ClearFlags.ZBuffer, _
        Color.Black.ToArgb(), 1.0, 0)

    DX9.BeginScene()
    DX9.Material = Material

    Mesh.DrawSubset(0)

    DX9.EndScene()
    DX9.Present()
End Sub
```

If you were to run the application at this time, you would see a scene rendered something like Figure 29.2.

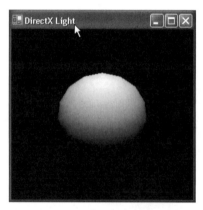

FIGURE 29.2 The scene rendered in DirectX is similar to its OpenGL counterpart.

SUMMARY

In this chapter, we built two quick examples for 3D rendering using two different APIs, both of which are extremely popular and can be very productive. The decision on which to use is dictated by your own preferences. In Chapter 30, Using Third-Party Engines, we're going to use a 3D engine instead of creating one with OpenGL or DirectX.

30 Using Third-Party Engines

In this chapter, we look at several of the third-party engines that are available to us for the development of applications capable of 3D rendering. This list is not all-inclusive, but does give you an idea of what is available. There are also many additional third-party development tools, such as Darkbasic, Director, Jamagic, and the 3D Game Studio to name a few. We only concern ourselves with the development tools that are COM-based or can be called from within a VB .NET application.

ON THE CD

The source code for the projects are located on the CD-ROM in the PROJECTS folder. You can either type them in as you go or you can copy the projects from the CD-ROM to your hard drive for editing.

3DLINX

The first of the 3D engines we look at is called 3DLinX (*www.3dlinx.com*). This is one of the most interesting of the offerings because it includes 3D models that are called Living Models™. They are aware of each other and interact with one another, obeying the laws of physics with very little programming. 3DLinX is a COM-based engine and a native .NET version is currently under development.

TRUEVISION3D

Truevision3D (*www.truevision3dsdk.com*) probably produces the best rendering of any of the engines that we are looking at. It is supported in VB .NET, C#, VB5, VB6, and Delphi. The engine is based on DirectX 8.1 and includes the ability to render terrains or you can use BSPs for inside environments. It also offers the most support for graphics with popular formats such as 3DS, X, MDL, MD2, and MD3. The only real negative to this engine is that it seems to have a problem with the digitizer of some of the Tablet PCs it has been tested on. This makes it a problem in using it for Tablet PC development and is the only reason that it was not the engine of choice.

REVOLUTION3D

If there is such a thing as engine "rivalries," then this is definitely the rival of Truevision3D. Revolution3D (*www.revolution.de*) is based on DirectX 9 and supports the same development environments as Truevision. It also offers support for many graphics formats, including Milkshape 3D, which makes it a popular choice for those on a budget. The future of this engine is very promising and it remains to be seen if either of the rival engines will ever outdistance themselves from one another.

3D STATE

3D State (*www.3dstate.com*) is the 3D engine formerly known as Morfit. It is a DLL file that includes over 200 graphics functions that you can call from VB (or many other languages). It supports the Quake MD2 format for animated 3D models and includes a variety of lighting effects and a relatively fast rendering engine. It's another good engine worth a look.

WILDTANGENT

WildTangent's Web Driver (*www.wildtangent.com*) platform allows games to be played online and is built on DirectX. One of the big advantages of WildTangent is the GameChannel™ delivery method, which allows games to be downloaded in the background with idle bandwidth. This delivery method is unparalleled and can be a good way for developers to get their software out to the masses. The 3D engine is probably not as feature-rich as some of the others, but it is more than adequate for what would normally constitute a Tablet PC game.

6DX

6DX (*www.aztica.com*) is the engine that we are going to use to develop a sample in this chapter. It is a general-purpose engine that can have a multitude of uses and is very easy to use for development. Before moving forward, it is probably a good idea to download the SDK, which you will find at the Web site and follow its installation instructions carefully.

We begin the development project by looking at creating a new Windows Forms application. Next, we need to add the control to our project. Right-click on the Toolbox, and from its context menu, you can choose Customize. Make sure the COM tab is selected in the Customize Toolbox window and try to locate the CA6DX control. If you see it listed, you can select it and click OK. Otherwise, you should browse to the directory where you saved/registered the control and choose the 6DX_atl.dll file from there. Either way, it is now visible in the Toolbox.

You can now add it to the form just like any other control. You can draw a box in the form to fill most of the center portion of the form (see Figure 30.1) and change its Name property to "C6DX".

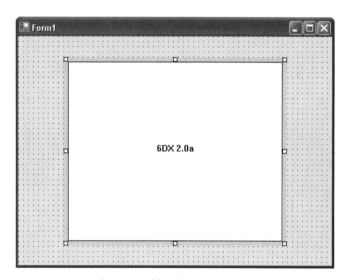

FIGURE 30.1   The control has been drawn on the form.

We are now going to add six Label controls to the form. They are arranged as in Figure 30.2 and have the following properties set as shown in Table 30.1 (the Tex-tAlign property should be set to Middle Center for all of them):

TABLE 30.1 Adding Label controls and setting properties

Name	Text
lblUp	∧
lblDown	∨
lblLeft	<
lblRight	>
lblCamUp	// \\
lblCamDown	\\ //

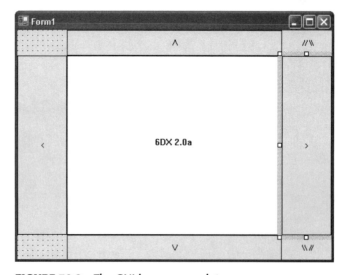

FIGURE 30.2 The GUI is now complete.

The final item we need to add to the form is a Timer control, and set the In-terval property to 100. The next step is to open the Code Editor, and add the following variables for our project:

```
Dim map As My6dx_atl.CA6DXMAP
Dim Act As My6dx_atl.CA6DXACTOR
Dim Mdl As My6dx_atl.CA6DXMODEL
```

```
Dim MdlInst As My6dx_atl.CA6DXMD3INSTANCE
Dim cam As New My6dx_atl.CA6DXCAMERA()
Dim Mover As New My6dx_atl.CA6DXCOLLIDER()
Dim MoveUp As Boolean
Dim MoveDown As Boolean
Dim RotateLeft As Boolean
Dim RotateRight As Boolean
```

Now, in the Form_Load event, we'll initialize the 6DX control, load a map and an actor, assign the camera, and then make the map visible. Here is the code:

```
Private Sub Form1_Load(ByVal sender As Object, ByVal e As
System.EventArgs) Handles MyBase.Load
    ChDir("c:\6dx\simple_runtime")
    Me.Text = "Tablet 3D Rendering"

    C6DX.Initialize(Handle.ToInt32, 1, "demo", "demo", "demo")

    map = C6DX.World.CreateMapInstance("doortest.map")
    Act = New My6dx_atl.CA6DXACTOR()
    Act.Create(True)
    Act.Visible = True

    Mdl = New My6dx_atl.CA6DXMODEL()
    Mdl.Create("md3\peasant.md3")
    MdlInst = Act.Attach(Mdl, "")
    Mover.AssignCamera(cam)
    map.Visible = True
End Sub
```

The previous code assumes that you installed the code to the C:\6DX directory. You need to change this hard-coded location if you installed it elsewhere.

We are now going to set up the Timer_Elapsed procedure, which contains the code for movement, updating the frame, rendering the frame, and showing the frame. The rendering, updating, and movement are all methods of the 6DX control, so they are easy to implement. For the movement, we use a series of If...Then statements to test for two conditions: key presses on a keyboard or mouse position over the Label controls that are visible outside of the control on the form. We test the Boolean values for MoveUp, MoveDown, MoveLeft, and MoveRight. We use two methods for movement so that a user with only their Tablet PC could explore just as easily as a user with a keyboard.

Here is the code:

```
Private Sub Timer1_Elapsed(ByVal sender As System.Object, ByVal e As
System.Timers.ElapsedEventArgs) Handles Timer1.Elapsed
    Static forward, fturn, ftime As Single

    ftime = C6DX.Time
    forward = 0
    fturn = 0

    If C6DX.IsKeyPressed(Keys.Up) Then forward = 128
    If C6DX.IsKeyPressed(Keys.Down) Then forward = -128
    If C6DX.IsKeyPressed(Keys.Left) Then fturn = -128
    If C6DX.IsKeyPressed(Keys.Right) Then fturn = 128

    If MoveUp Then forward = 128
    If MoveDown Then forward = -128
    If RotateLeft Then fturn = -128
    If RotateRight Then fturn = 128

    If C6DX.IsKeyPressed(Keys.Escape) Then
        Timer1.Enabled = False
    End If

    Mover.CalcCharacterMovement(forward, 0, fturn, ftime)
    Mover.DoCollision()

    C6DX.UpdateFrame(C6DX.Time)
    C6DX.RenderFrame()
    C6DX.ShowFrame()
End Sub
```

We know that the Boolean values are tested and used for movement. We assign
their values by using the MouseEnter and MouseLeave events for the various labels.
Inside the events, we also change the foreground colors for the labels so that the
user can tell when their mouse is positioned over the various movement options.
Here is the code for all of the MouseEnter and MouseLeave events:

```
Private Sub lblUp_MouseEnter(ByVal sender As Object, ByVal e As
System.EventArgs) Handles lblUp.MouseEnter
    lblUp.ForeColor = Color.Red
    MoveUp = True
End Sub

Private Sub lblUp_MouseLeave(ByVal sender As Object, ByVal e As
System.EventArgs) Handles lblUp.MouseLeave
    lblUp.ForeColor = Color.Black
    MoveUp = False
```

```
End Sub

Private Sub lblDown_MouseEnter(ByVal sender As Object, ByVal e As
System.EventArgs) Handles lblDown.MouseEnter
    lblDown.ForeColor = Color.Red
    MoveDown = True
End Sub

Private Sub lblDown_MouseLeave(ByVal sender As Object, ByVal e As
System.EventArgs) Handles lblDown.MouseLeave
    lblDown.ForeColor = Color.Black
    MoveDown = False
End Sub

Private Sub lblLeft_MouseEnter(ByVal sender As Object, ByVal e As
System.EventArgs) Handles lblLeft.MouseEnter
    lblLeft.ForeColor = Color.Red
    RotateLeft = True
End Sub

Private Sub lblLeft_MouseLeave(ByVal sender As Object, ByVal e As
System.EventArgs) Handles lblLeft.MouseLeave
    lblLeft.ForeColor = Color.Black
    RotateLeft = False
End Sub

Private Sub lblRight_MouseEnter(ByVal sender As Object, ByVal e As
System.EventArgs) Handles lblRight.MouseEnter
    lblRight.ForeColor = Color.Red
    RotateRight = True
End Sub

Private Sub lblRight_MouseLeave(ByVal sender As Object, ByVal e As
System.EventArgs) Handles lblRight.MouseLeave
    lblRight.ForeColor = Color.Black
    RotateRight = False
End Sub
```

There are still two Label controls that we haven't dealt with. We use the `Click` events for both labels. Inside of the events, we set the pitch of the camera up five degrees or down five degrees depending on which label is clicked.

Here is the code:

```
Private Sub lblCamUp_Click(ByVal sender As Object, ByVal e As
System.EventArgs) Handles lblCamUp.Click
    Dim x As My6dx_atl.dxx_angle
```

```
    x = cam.Orientation
    x.pitch = cam.Orientation.pitch - 5
    cam.Orientation = x
End Sub

Private Sub lblCamDown_Click(ByVal sender As Object, ByVal e As
System.EventArgs) Handles lblCamDown.Click
    Dim x As My6dx_atl.dxx_angle
    x = cam.Orientation
    x.pitch = cam.Orientation.pitch + 5
    cam.Orientation = x
End Sub
```

The application is now finished and can be saved. You can also test it to see if the rendering works correctly and that the movement works as well.

SUMMARY

In this chapter, we built an application that loads a precompiled map and allows us to move around it. We used a third-party engine called 6DX, which has a demo version available for download. In the final chapter of the book, we build a faxing and e-mailing program.

31 Tablet Fax

In the final chapter of the book, we develop an application that allows us to send faxes. We also use factoids, which improve accuracy for specific types of pen-based input. To send the fax, we use the Faxcom.dll, which should be available on all Tablet PCs

The source code for the projects are located on the CD-ROM in the PROJECTS folder. You can either type them in as you go or you can copy the projects from the CD-ROM to your hard drive for editing.

BEGINNING THE PROJECT

To begin the project, create a new Windows Forms application and add the controls shown in Table 31.1, using Figure 31.1 as a reference.

TABLE 31.1 Adding controls to the application

Type	Name	Text
TextBox	FaxTo	empty
TextBox	FaxNum	empty
TextBox	FaxSubject	empty
TextBox	FaxNote	empty
CheckBox	CoverPage	Cover Page
CheckBox	CheckBox1	Display as Ink
InkEdit	InkEdit1	InkEdit1

FIGURE 31.1 The GUI is finished.

We are now set to add a reference to `Microsoft.Ink` and `Microsoft.Ink.15`. You also need a reference to the Microsoft Fax Service Extended COM Type Library, which can be found on the COM tab.

The next step is to add the `Imports` for `Microsoft.Ink`:

```
Imports Microsoft.Ink
```

We use a Pen Input Panel, so we can create this and also handle disposing of it on the `Form_Closing` event:

```
Dim thePenInputPanel As New PenInputPanel()


Private Sub Form1_Closing(ByVal sender As Object, ByVal e As
System.ComponentModel.CancelEventArgs) Handles MyBase.Closing
    thePenInputPanel.Dispose()
End Sub
```

We now deal with the Pen Input Panel and using factoids in the TextBox controls, which includes phone numbers, and so on. Factoids greatly improve recognition (much like our custom grammar for SAPI) by providing the recognizer with a definition of the words, numbers, and punctuation that are expected. Factoids are available for items such as phone numbers and e-mail addresses.

We are able to assign the factoid and the Pen Input Panel in the respective TextBox_Enter events:

```
Private Sub FaxTo_Enter(ByVal sender As System.Object, ByVal e As
System.EventArgs)
    thePenInputPanel.AttachedEditControl = FaxTo
    thePenInputPanel.Factoid = Factoid.Default
End Sub

Private Sub FaxNum_Enter(ByVal sender As System.Object, ByVal e As
System.EventArgs)
    thePenInputPanel.AttachedEditControl = FaxNum
    thePenInputPanel.Factoid = Factoid.Telephone
End Sub

Private Sub FaxSubject_Enter(ByVal sender As System.Object, ByVal e
As System.EventArgs)
    thePenInputPanel.AttachedEditControl = FaxSubject
    thePenInputPanel.Factoid = Factoid.Default
End Sub

Private Sub FaxNote_Enter(ByVal sender As System.Object, ByVal e As
System.EventArgs)
    thePenInputPanel.AttachedEditControl = FaxNote
    thePenInputPanel.Factoid = Factoid.Default
End Sub
```

Let's now handle the changing of the InkEdit control to display the information as ink or as the recognized text. We use CheckBox1 for this. The other Sub procedure that deals with the way the InkEdit control works is the Form_Load event, which empties any text inside the control and sets its Width property to 100.

Here are the two procedures:

```
Private Sub CheckBox1_CheckedChanged(ByVal sender As System.Object,
ByVal e As System.EventArgs) Handles CheckBox1.CheckedChanged
    InkEdit1.SelectAll()
    If CheckBox1.Checked = True Then
        InkEdit1.SelInksDisplayMode = InkDisplayMode.Ink
    Else
```

```
            InkEdit1.SelInksDisplayMode = InkDisplayMode.Text
        End If
End Sub

Private Sub Form1_Load(ByVal sender As System.Object, ByVal e As
System.EventArgs) Handles MyBase.Load
    InkEdit1.Text = ""
    InkEdit1.DrawingAttributes.Width = 100
End Sub
```

The final Sub procedure is the longest one. It occurs when btnFax is clicked. It begins by creating a new fax document, fax server, sender, and job ID:

```
Dim objFaxDocument As New FAXCOMEXLib.FaxDocument()
Dim objFaxServer As New FAXCOMEXLib.FaxServer()
Dim objSender As FAXCOMEXLib.FaxSender
Dim JobID As Object
```

The Sub procedure continues on with the following steps:

1. Connect to the local fax server.
2. Set the body of the fax document to the content of InkEdit1.
3. Use the FaxSubject Text property to set the document name.
4. Set the priority to high.
5. Add the recipient name and fax number.
6. Attach the fax to the fax receipt.
7. Add a cover page if the CoverPage check box is selected.
8. Set the Subject equal to the FaxSubject's Text property.
9. Set the JobID.
10. Display a message box with the ID.
11. The entire set of steps have simple error handling and an error is displayed if anything happens.

Here is the code for the entire procedure:

```
    Private Sub btnFax_Click(ByVal sender As System.Object, ByVal e As
System.EventArgs)
        Dim objFaxDocument As New FAXCOMEXLib.FaxDocument()
        Dim objFaxServer As New FAXCOMEXLib.FaxServer()
        Dim objSender As FAXCOMEXLib.FaxSender
        Dim JobID As Object

        On Error GoTo Error_Handler
```

```
                    objFaxServer.Connect("")
                    objFaxDocument.Body = InkEdit1.Rtf
                    objFaxDocument.DocumentName = FaxSubject.Text
                    objFaxDocument.Priority = FAXCOMEXLib.FAX_PRIORITY_TYPE_
ENUM.fptHIGH

                    objFaxDocument.Recipients.Add(FaxNum.Text, FaxTo.Text)

            objFaxDocument.AttachFaxToReceipt = True

            If CoverPage.Checked Then
                objFaxDocument.CoverPageType = FAXCOMEXLib.FAX_COVERPAGE_
TYPE_ENUM.fcptSERVER
                objFaxDocument.CoverPage = "generic"

                objFaxDocument.Note = FaxNote.Text
                objFaxDocument.Sender.LoadDefaultSender()
            End IF

            objFaxDocument.Subject = FaxSubject.Text

            JobID = objFaxDocument.ConnectedSubmit(objFaxServer)

            MsgBox("The Job ID is :" & JobID(0))
            Exit Sub

Error_Handler:
            MsgBox("Error : " & Hex(Err.Number) & ", " & Err.Description)

        End Sub
```

That's it for this final program. You can test it out to see if it is working correctly.

SUMMARY

Congratulations on making it through this entire text. You should now have a good understanding of the wide variety of applications that can be created for the Tablet PC. When learning any new programming concepts, you will undoubtedly have many questions. There are several great places to learn online, including the TabletPC Developer Web site at *www.tabletpcdeveloper.com* and the Tablet PC developer usenet newsgroup: *microsoft.public.windows.tabletpc.developer.*

APPENDIX ■ About the CD-ROM

The CD-ROM included with *Developing Tablet PC Applications* includes all of the necessary tools (with the exception of Visual Basic) to write the programs that are developed in each chapter. It also includes full color images of all the figures in the book, and the source code and executable files for the sample projects.

CD FOLDERS

On the CD-ROM, you'll find the following folders:

FIGURES: The full color version of all the figures in the book

PROJECTS: Arranged by chapter and includes the source code and executable files for every sample in the book

APPLICATIONS: We have included the full version of the DirectX 9.0b SDK along with everything you need to develop applications for Microsoft Agent.

SYSTEM REQUIREMENTS

The minimum system requirements are as follows:

■ Microsoft Visual Basic .NET
■ Windows 2000, XP
■ 450-megahertz (MHz) Pentium II-class processor, 600-MHz Pentium III-class processor recommended
■ CD-ROM
■ Hard Drive: 200MB of free space to install the examples
■ 32MB of RAM

INSTALLATION

To use the programs on the CD-ROM, your system should match at least the minimum system requirements. The image files are in TIFF format.

Index

* asterisk character, 218
[ ] brackets, 218
∧ carat character, 162
( ) parentheses, 218
% percentage sign character, 162
+ plus sign character, 162, 218
_ underline character, 180
| vertical bar character, 218–219

A

Abs method, 84–85
Absolute values, 84–85
Abstraction, 61–62
AccessibleDescription Managed Library
　　property, 144
AccessibleName Managed Library prop-
　　erty, 145
AccessibleRole Managed Library prop-
　　erty, 145
ActiveX controls, 125–126
　InkEdit properties available, 136–137,
　　138
ADO.NET, 309
Agent
　click events and, 219
　declaring character files for, 209
　defined and described, 205–206
　downloads, 207
　grammar options, special characters
　　for, 218
　initializing a character, 210–212

installation, 207–209
project tutorial, 219–223
speech recognition, 215–217
testing Agent projects, 212–213
Anchor Managed Library property, 145
APIs
　calls, 29–30, 180
　Component Object Model (COM)
　　APIs, 121
　data type conversions during upgrades,
　　29–30
　Ink API, 130
　Ink Data Management API, 130
　Ink Recognition API, 130
　Managed APIs in Software Develop-
　　ment Kit (SDK), 130
　.NET equivalents for API calls, 40–41,
　　187–191
　Pen API, 130
　Power Management API, 265–275
　Recognition API, 130
　SAPI (Speech API), 225–233
　Tablet Input API, 130
　in Tablet PC SDK, 121
Appearance property for ActiveX con-
　　trols, 136
Application templates, 93–94
Arrays
　control arrays, 53
　declaration of, 33, 51
　dynamic arrays, 52–53